SpringerBriefs in Applied Sciences and Technology

SpringerBriefs present concise summaries of cutting-edge research and practical applications across a wide spectrum of fields. Featuring compact volumes of 50–125 pages, the series covers a range of content from professional to academic.

Typical publications can be:

- A timely report of state-of-the art methods
- An introduction to or a manual for the application of mathematical or computer techniques
- A bridge between new research results, as published in journal articles
- A snapshot of a hot or emerging topic
- An in-depth case study
- A presentation of core concepts that students must understand in order to make independent contributions

SpringerBriefs are characterized by fast, global electronic dissemination, standard publishing contracts, standardized manuscript preparation and formatting guidelines, and expedited production schedules.

On the one hand, **SpringerBriefs in Applied Sciences and Technology** are devoted to the publication of fundamentals and applications within the different classical engineering disciplines as well as in interdisciplinary fields that recently emerged between these areas. On the other hand, as the boundary separating fundamental research and applied technology is more and more dissolving, this series is particularly open to trans-disciplinary topics between fundamental science and engineering.

Indexed by EI-Compendex, SCOPUS and Springerlink.

More information about this series at http://www.springer.com/series/8884

Fadzli Mohamed Nazri ·
Mohd Azrulfitri Mohd Yusof ·
Moustafa Kassem

Precast Segmental Box Girders

Experimental and Analytical Approaches

 Springer

Fadzli Mohamed Nazri
School of Civil Engineering
Universiti Sains Malaysia
Penang, Malaysia

Mohd Azrulfitri Mohd Yusof
Jambatan Kedua Sdn. Bhd.
Penang, Malaysia

Moustafa Kassem
Universiti Sains Malaysia
Penang, Malaysia

ISSN 2191-530X ISSN 2191-5318 (electronic)
SpringerBriefs in Applied Sciences and Technology
ISBN 978-3-030-11983-6 ISBN 978-3-030-11984-3 (eBook)
https://doi.org/10.1007/978-3-030-11984-3

Library of Congress Control Number: 2019930573

This Springer imprint is published by the registered company Springer Nature Switzerland AG
The registered company address is: Gewerbestrasse 11, 6330 Cham, Switzerland

Preface

This book explains the fundamental of elastic behaviour of erected precast segmental box girder (SBG) when subjected to static load and also the construction process (casting and erection work) involved. The SBG is the preferred super-structural element in bridge design and construction, especially in Malaysia which is home to the longest bridge in South-East Asia.

The objective of this book is to analyse and compare the elastic behaviour of erected precast SBG subjected to Static Load Test obtained experimentally, with (to?) the results obtained using the Finite Element Method (FEM) and theoretical calculations, under short-term deflection analysis for different loads. This is obtained by determining the maximum deflection, stress and strain value of single span precast SBG under a variety of transversal slope. The work presented herein is important to industries conducting static load test prior to inauguration the bridge to traffic and may be potentially of interest to researchers, designers and engineers looking to validate experimental work with numerical and analytical approaches.

This book consists of four chapters; Chap. 1 gives an overview of the construction process of the precast SBG from the assembly to the casting stage of Penang Second Bridge. Chapter 2 presents the previous researches and/or studies related to experimental and analytical work on segmental bridges. Chapter 3 describes the deck deflection observations by experimental and FEA method, which was subjected to static load. It then describes the parametric study on the models for a variety of transversal slopes, in addition to the standard guidelines related to load cases. Chapter 4 discusses and presents the outcome obtained from this research work.

We sincerely appreciate the financial support given by the Jambatan Kedua Sdn. Bhd. (JKSB) for giving permission and assistance in providing information, and data collection as well as scholarship approval from the Second Penang Bridge Project for this study.

Penang, Malaysia   Fadzli Mohamed Nazri
Penang, Malaysia   Mohd Azrulfitri Mohd Yusof
Penang, Malaysia   Moustafa Kassem

Contents

1 Description of SBG Assembling and Casting-Penang Bridge 1
  1.1 Introduction 1
  1.2 Bridge Description 1
  1.3 Segment Casting..................................... 4
  1.4 Short-Line Method 5
  1.5 Segment Transportation 7
  1.6 Segment Erection 11
  References 13

2 Overview of Precast Segmental Box Girder 15
  2.1 Introduction 15
  2.2 Previous Study for Segmental Bridge Under Static Load Test 16
  2.3 Elastic Behaviour, Displacement, Stress and Strain
       of Segmental Bridges 19
  2.4 Previous Finite Element Analysis (FEA) Studies of Segmental
       Bridges 20
  2.5 Previous Transversal Slope Studies of Segmental Bridges 25
  2.6 Summary of Previous Studies 26
  References 28

3 Finite Element Analysis of SBG Subjected to Static Loads 31
  3.1 Materials and Properties of Precast SBG 31
  3.2 Load Test Execution.................................. 35
  3.3 Strain Gauge Measurement 36
  3.4 Finite Element Analysis 37
      3.4.1 Description of Finite Element Model 38
      3.4.2 Accuracy of Finite Element Analysis 39
      3.4.3 Attributes for Modelling.......................... 40

3.4.4   Comparison with Deflection   42
3.4.5   Comparison with Strain   45
3.5   Summary of Methodology   46
References   46

4   Validation of Experimental and Analytical Study Work   49
4.1   Elastic Behaviour of Precast SBG   49
4.2   Comparison with Deflection   51
4.3   Parametric Study for Different Transversal Slope Under
Maximum Deflection Along the Span Length of the Precast
SBG   57
4.3.1   Stress and Strain Analysis at Mid-Span
for 2.5% Transversal Slope   59
4.4   Chapter Outcome Summary   69
References   74

Symbols, Acronyms and Abbreviations

1D	One Dimensional
2D	Two Dimensional
3D	Three Dimensional
DA	Deflection Analysis
DIM	Double Integration Method
FDOT	Florida Department of Transportation
FEA	Finite Element Analysis
FEM	Finite Element Method
GPS	Global Positioning System
JKSB	Jambatan Kedua Sdn. Bhd.
KEL	Knife Edge Load
LB	Longitudinal Bending
LBM	Line Beam Method
LC	Load Case 1
LC2	Load Case 2
LC3	Load Case 3
LRT	Light Rail Transit
MRT	Mass Rapid Transit
NDT	Non-Destructive Test
QA	Quality Assurance
QC	Quality Control
SBG	Segmental Box Girder
SEP	Segmental External Prestressed
SHMS	Structural Health Monitoring System
SLS	Serviceability Limit State
TB	Transverse Bending
UDL	Uniformly Distributed Load
ULS	Ultimate Limit State

UPV	Ultrasonic Pulse Velocity
VB	Vertical Bending
VWSW	Vibrating Wire Strain Gauge

Chapter 1
Description of SBG Assembling and Casting-Penang Bridge

Abstract This chapter presents a brief description on precast segmental box girders technology of the longest South-East Asia Bridge in Penang, Malaysia followed by the construction process in terms of SBG Casting, SBG Transporting, and SBG Erecting. It highlights the importance of performing experimental static load test in construction of segmental bridges before it can be opened to the public. This chapter also notes the comparison between the period of assembly of the bridge superstructure to the cast-in-place construction.

1.1 Introduction

The Second Penang Bridge is part of a traffic logistics connecting Batu Kawan and Batu Maung across the Southern Channel of the Penang Straits. The total length of the highway is approximately 24 km and is set to be the longest bridge in South-East Asia as shown in Fig. 1.1. The deck of the bridge is approximately 16.9 km over water consists of 4-lane dual carriageway and 2-lane dedicated motorcycle. The structural type of this bridge is a cable-stayed bridge that has a central span of 475 m and 584 numbers of piers respectively. Package 3 comprises mainly the expressway and has been awarded to various local contractors. The construction of precast segmental box girder is commenced in July 2010 and completed in December 2012 while for the launching works is completed in February 2013. The Second Penang Bridge Project was opened to the public in 1st of March 2014. The logistical, quality, environmental, safety and health factors were the challenges in this project specifically for the superstructure works activities.

1.2 Bridge Description

The Second Penang Bridge superstructure consists of 289 spans on nominal length 55 m side by side trapezoidal box girder and contains 14 segments in total which

Fig. 1.1 Second Penang Bridge project alignment

are Typical Segments; 9 numbers, Deviator Segments; 2 numbers, Pier or Abutment Segment; 2 numbers and Locator; 1 number. The segments width is varying from 2.65 to 4.1 m and nominal 3.2 m deep. The bridge designed with dual carriageway and dedicated motorcycle lane is shown in Fig. 1.2. The segments, each weighing from 69 to 100 tonne, were cast by BCSB JV Persis Sdn. Bhd in its 3 casting yards (Open Casting Yard 1, Open Casting Yard 2 and Covered Casting Yard) [1]. In order to employ local materials and skills manpower as much as possible, as well as the desire to minimise cast in situ works over the sea, the most economical span length and suitable form of construction was chosen [2]. Two viaducts modules of 6 or 5 spans continuous between expansion joints were adopted for the approach spans, minimizing the number of expansion joints at every 275 and 330 m respectively and eventually for future maintenance. The deck uses epoxy glue joint system with combination of internal and external pre-stressing.

Analysis and design for the Second Penang Bridge Project was carried out in accordance with the project relevant specification to satisfy loading requirements under service and ultimate limit state. The typical bridge units were also supported on the High Damping Rubber Bearing (HDRB) and being the longest of HDRB installation for marine segmental bridge in the world. HDRB has the ability to withstand large displacement in bilateral and rotational direction, durable with minimal maintenance as well as utilizing natural rubber available locally as shown in Fig. 1.3 [6].

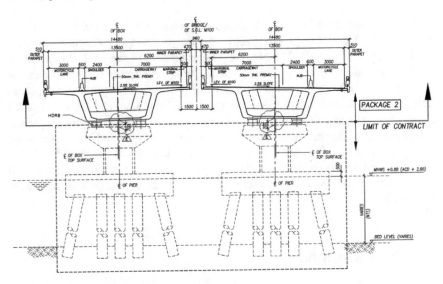

Fig. 1.2 Bridge sectional view

Fig. 1.3 HDRB with and without preset plate at expansion joint area underneath abutment segments

The materials used in the precast segment production are required to comply with the Project Specification requirements. Design grade of concrete is specified as Grade 55 (55 N/mm^2). The nominal binder content of the concrete should be between 380 and 450 kg/m^3 including silica fume. An approved superplasticizer is incorporated into the concrete mix. Most of the concrete raw materials such as Pulverised Fly Ash Cement (PFA), fine aggregate, 20 mm graded aggregate, silica fume, concrete admixtures, and reinforcement bars are supplied directly by local companies. The concrete used in the segments includes fly ash and high-range water reducing admixture, allowing it to develop high early strength. The targeted design strength is 67.5 MPa while the early strength is 15 MPa for mould striking and 25 MPa for lifting. The current average strengths for 28 days is over 90 MPa were achievable.

1.3   Segment Casting

The segments were fabricated in a casting yard that was previously an agricultural pineapple plantation at Batu Kawan and covers an area of 48 acres, see Fig. 1.4. Segments are manufactured in the precast yard to a casting schedule that is largely determined by the proposed erection programme. The casting yard was chosen strategically location to facilitate easy transportation of all the segments by barges. In this project, a total of 8092 segments (Concrete volume: 260,000 m^3 and steel reinforcement: 60,000 tonne) were fabricated in a period of 30 months [3]. The reinforcement bars of varying diameters from T10 to T40 are tie-wired together to form a reinforcement cage. Prefabrication of the reinforcement cages are done in the jigs that define the segment concrete shape less cover, and some cast-in items shall be incorporated during the rebar fixing, as it does not cause rebar cages distortion during lifting and handling. These cast in-items includes internal tendon corrugated PVC, deviator tubes, drainage gully etc.

Due to the importance of the project, 22 sets (6 pier or abutment moulds and 16 typical moulds) of short-line casting moulds or cells were produced on average 15 numbers per day. The actual production characteristics, however reached a typical cycle of 1.5 day for the typical or standard segments and 3 days for the pier and abutment segments respectively after a brief initial learning curve period of 2 months. The casting yard were equipped with 2 numbers of 120 tonne heavy gantry cranes, 3 numbers of straddle carriers for segment handling and transfer within their respective storage areas and 5–20 tonne overhead gantry cranes for light duty task such as manipulating the reinforcement cages and concreting works. The casting yard had a storage capacity of 1000 segments based on double stacking see Fig. 1.5. Concrete was mixed in the casting yard by using 2 numbers of batching plants of 120 m^3/hr capacity each and was operated by Unipati Concrete (UPC) with associated material storage.

Fig. 1.4 Batu Kawan
casting yard

Fig. 1.5 Segments storage
(double stacking)

1.4   Short-Line Method

The short-line casting method was selected because it does not require extensive casting facilities, special heavy lifting equipment and storage as shown in Fig. 1.6 [6]. It was not a new in Malaysia and had always been practiced in several projects locally such as Sungai Prai Bridge Project in Penang, Kuala Lumpur LRT and Malaysia MRT Project. To ensure that the segments fit together when assembled in their final position, the concept of "match-casting" is employed. Match casting is the technique of casting a new segment between a fixed form on one end and its neighbouring segment on the other end [5]. Spans are cast and erected as a simply supported span by span method arrangement and then joined by in situ stitches at piers from continuous bridge modules between the expansion or movement joints.

Fig. 1.6 Match-casting
using short-line method [5]

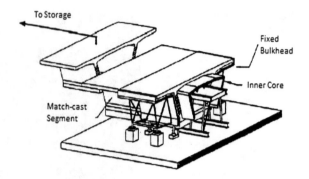

Generally, the spans are arranged in modules of six spans, with the last five modules at the mainland abutment being five spans.

The two pier segments are cast in a pier mould by match-casting against the adjacent span segment Fig. 1.7. In the short-line method, each segment is cast and subsequently moves into "match-cast" position before pouring the next segment. Placement of the match-cast segment is of primary concern to achieve the theoretical geometry which includes cambering for expected structural displacement. The accuracy of calculations and proper control of relative placement in the forms will greatly determine the degree of success of the erection process and the final geometry of the constructed precast segmental bridge. The success of short-line match-cast joint method relies heavily on accurate geometry control during match casting as the scale of adjustments during erection is very small and difficult to implement. The required levels of accuracy in positioning the segments match-cast against each other are stringent in order to assure acceptable tolerances in the geometry of the structure [4]. As a result, the segments were cast in the order in which they were erected. The reason for this precision was the exactly quality control demanded by the owner of this project. Tolerances between segment surveys were to be between 0.3 and 0.5 mm. The bridge deck alignment is produced by alignment control, which involves precise survey in the order of a fraction of a millimetre for line, levelling and setting up. This is controlled by using special alignment control software with alignment data provided by the specialist consultant. The casting sequences of SBG are shown in Fig. 1.8.

Remedial works for the more severe cases improved with improved procedures, supervision and application methods. These works have been subjected to Ultrasound Pulse Velocity (UPV) testing and MP4 accelerometer testing in addition to the standard cube strength (Compression Test), Rapid Chloride Permeability Test (RCPT) and Initial Surface Absorption Test (ISAT) to confirm the structural integrity and durability compliance as shown in Figs. 1.9 and 1.10.

The Second Penang Bridge proved that assembly of the bridge superstructure takes much less time than cast-in-place construction, as precast segments have gained more strength and do not need to cure on site before being pre-stressed together. The fundamental process to address in casting productivity for Typical or Pier segments

Fig. 1.7 Principle activities
involved in short-line
match-casting operations

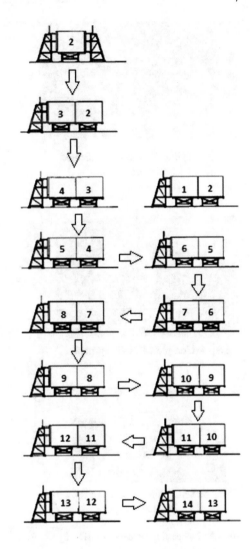

is the cycle time of the moulds. With a fixed number of moulds being available, the
faster each mould can be recycled, the higher the productivity [3].

1.5 Segment Transportation

Figure 1.11 shown the completed SBGs are transported to the storage yard by the
straddle carrier and 120 tonne heavy gantry crane. The segments can only be lifted
from the soffit table upon the concrete achieving its minimum compressive strength
of 25 MPa in accordance to project specification. Straddle carrier will move over

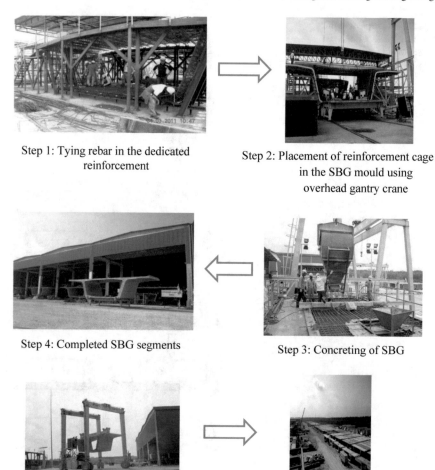

Step 1: Tying rebar in the dedicated reinforcement

Step 2: Placement of reinforcement cage in the SBG mould using overhead gantry crane

Step 4: Completed SBG segments

Step 3: Concreting of SBG

Step 5: SBG transported to yard by Straddle Carrier

Step 6: SBG storage yard

Fig. 1.8 The casting sequence of SBG by short-line method

the segment to be lifted and its lifting beam shall be lowered into the segment. The lifting frame shall be connected to the segment using 4 numbers of 40 mm diameter pre-stress high tensile bars. The stress bar shall be stressed to the required force. Once stressing is completed, the segment can be lifted safely. For the quality inspection purposes, the segment is then lifted from the soffit table to check of any defects to the concrete soffit such as grout loss, honeycombing or cracking. After completion of inspection, the straddle carrier then shall lift and move the segment to the appropriate heavy gantry crane transfer bay in the storage yards. At the storage yard, a heavy gantry crane is used to lift, move and placing the segments on the 3

Fig. 1.9 RCPT and ISAT
tests

Fig. 1.10 Initial surface
absorption test

pedestal supports to avoid warping effects on the segments. The same practice is
also applied to double stacking of the segment.

Precaster UEMB must to make sure that stressing bars reading at 5000 psi
(350 kg/cm^2) before lifting. As a preventive action, the stressing bars and embedded
couplers need to change every 2500 numbers production of SBG in order to avoid
any mishandling or segment damages. Before transporting the segments to the jetty,
the segments lifting beams which are stored at the storage yard (after being used
for lifting at the erection fronts) shall be fixed to each segment. The beam shall be
connected (anchored) to the segment by using 4 numbers of 36 mm diameter prestressed bars which each one pre-stressed at 580 kN. Segments are delivered from
the jetty to the segment delivery barge using 3 numbers of 120 tonne gantry cranes
which is founded on dedicated rails at the jetty (see Fig. 1.12). Each barge can hold
up to 5 numbers segments and has a plan dimension of 120 × 50 ft (see Fig. 1.13).

Fig. 1.11 Segment being moved to storage by straddle carrier and heavy gantry crane

Fig. 1.12 Load out jetty to transport SBG from the casting yard to site

Fig. 1.13 Segments delivery by barge brought into position and attached to the working barge

Fig. 1.14 LG 1 and LG 2 in operation

1.6 Segment Erection

The total erection period was about 24 months. Since the Second Penang Bridge to be the longest bridge in South-East Asia, the aesthetics of approach structure was an important aspect of the design. The precast segments were erected using 4 numbers of Overhead Launching Girders namely LG1, LG2, LG3 and LG4 as shown in Fig. 1.14, designed, fabricated and supplied by DEAL, Italian Company. The segment decks uses epoxy glue joint system were longitudinally post-tensioned at site internally and externally under first stage and second stage stressing. Thus, shear keys cast into the segments guaranteed a perfect fit. As the gantries placed each span, the segments were temporarily secured with post-tensioning bars until the final post-tensioning tendons were installed and stressed.

The 128 m long LGs weighted about 600 tonne each and functioned both somewhat as a cantilever and span by span girder at the same time. Each girder had 1 service crane and 1 winch gantry with lifting capacities of 50 and 280 kN respectively. The lifting height of the winches was allowed for up to 30 m so that the

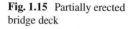
Fig. 1.15 Partially erected
bridge deck

segments could be picked up at sea level. The LGs had been designed to resist a
maximum wind speed of 97 km/hr. However, the maximum wind speed was limited to approximately 40 km/hr during self-launching and 70 km/hr during segment
erection.

The tendons are stressed into two stages. The first is to take the self-weight simply
supported between piers and the second stage continuity pre-stress is applied once
the stitches between pier segments are completed. External pre-stressing tendons
used are 31T15 of which there are 4 pairs anchored at the abutment and internal
tendons are 19T15 units for which there are two pairs anchored at the abutment.
Multistrand pre-stressing jacks are used to perform group stressing means that all
strands in one tendon are stressed together in a load-controlled manner and in incre-
ments. The spans are erected simply supported initially and then stitches are carried
out between pier segments to create continuity. The abutment segments are placed
on temporary bearings in the temporary stages before later being to the permanent bearings (HDRB) (see Fig. 1.15). Using segmental construction with overhead
launching girders allowed many construction activities to be completed away from
the site especially for the Second Penang Bridge project All pre-stressing tendons,
whether internal or external are grouted with cement under pressure. The purpose is
to ensure that the tendons are protected against chemical ingress that would cause
rusting of highly stressed strands. High quality requirements are imposed on the
wet and hardened cement to ensure a dense and homogenous cement material is
formed within the High-density Polyethylene (HDPE) external of internal corrugated ducts. Cement grout is composed of cement, water and admixture. Admixtures
include water-reducers and superplasticizers. The SBG erection sequence is shown
in Fig. 1.16.

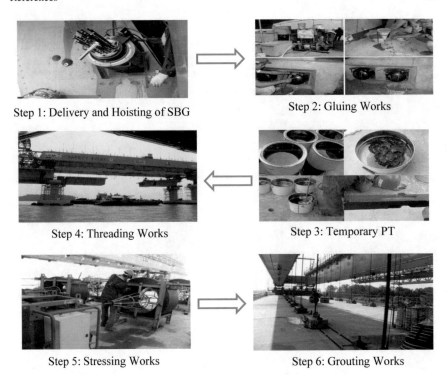

Fig. 1.16 The SBG erection sequence using span by span method

References

1. J. Clarke, in *Quality of Segmental Box Girder—Solving Issues During Construction*. 2nd International Seminar on the Design and Construction of the Second Penang Bridge. (Jambatan Kedua Ptd. Ltd., Sunway Putra Hotel, Kuala Lumpur, Malaysia, 2012)
2. A. Forouzani, *Penang Second Crossing Package 2: Box Girder Decks*. Technical Paper 4. (Jambatan Kedua Ptd. Ltd, Malaysia, 2011)
3. P.D. Kanapathy, in *Project Planning for SBG Production and Launching Works*. 2nd International Seminar on the Design and Construction of the Second Penang Bridge. (Jambatan Kedua Ptd. Ltd., Sunway Putra Hotel, Kuala Lumpur, Malaysia, 2012)
4. K. Kumar, K. Varghese, K.S. Nathan, K. Ananthanarayanan, in *Automated Geometry Control of Precast Segmental Bridges*. The 25th International Symposium on Automation and Robotics in Construction, vol. 26. (2008)
5. B. Levintov, Construction equipment for concrete box girder bridges. Concr. Int. **17**(2), 43–47 (1995)
6. M.I. Taib, *PB2X Construction Challenges—Second Penang Bridge. The Ingenieur*. (Board of Engineers, Malaysia, 2013)

Chapter 2
Overview of Precast Segmental Box Girder

Abstract This chapter gives updates on the current work for segmental box girder (SBG) under static load test and the measurement to determine the elastic behavior, displacement, stress and strain of the SBG. Moreover, study on finite element analysis (FEA) and transversal slope on SBG is also highlighted in this chapter. At the end of this chapter, a summary of all experimental and analytical research of SBG previously conducted, is given.

2.1 Introduction

The popularity of Precast segmental box girder (SBG) bridge construction has grown worldwide in the last few decades and has become a very popular method of bridges construction in Malaysia. This type of bridges offers many benefits to owners especially the Malaysian Government to help reduce cost, construction time, environmental impact, and maintenance of traffic. These benefits can be achieved while utilizing local labour and materials, better quality control and with minimum requirements for future maintenance. Furthermore, this method is also offers additional structural advantage of durability, fire resistance, deflection control, better ride serviceability, insensitivity to fatigue and inherent aesthetic appeal over the past 30 years. According to Rombach [1], the precast concrete segmental hollow box girder bridges externally prestressed are the preferred structure for many great elevated highways especially in South-East Asia.

Previous researches that are related to SBG will be discussed further in this chapter. There were many researches that were related to SBGs, however each of them differs from one to another in term of case studies and also the scope of work involved. After reading and researching previous studies, they can be divided into three which are experimental, numerical and a combination of both methods. Jazlan [2] reported that the trend emerged in Malaysia when the first segmental bridge was being constructed for the Light Rail Transit (LRT) in Kuala Lumpur and followed by the Second Link to Singapore in Johor and Ampang—KL Elevated Highway. It is not a new method in Malaysia and has always been practiced in several projects

© The Author(s), under exclusive license to Springer Nature Switzerland AG 2019
F. Mohamed Nazri et al., *Precast Segmental Box Girders*,
SpringerBriefs in Applied Sciences and Technology,
https://doi.org/10.1007/978-3-030-11984-3_2

locally such as the Sungai Prai Bridge Project in Penang, the Second Penang Bridge Project and also presently being used in Mass Rapid Transit (MRT) and LRT Project in Kuala Lumpur and Klang Valley.

The static load test is frequently used before the opening of a new large bridge to verify the actual structural behaviour of the bridge compared with that predicted by theory or conventional method. In this chapter, various types and construction methods of segmental bridges have been studied. However, the testing and modelling of precast SBG was not as common and it is a complex and tricky work. There have been numerous studies that examine the behaviour of SBG bridges. For example, Navratil and Zich [3] in their studies mentioned the possible reasons for excessive bridge deflections. It is often encountered in practice that the long-term deflections of prestressed bridges are greater than the deflections expected in design. The explanation of possible reasons for such large deflections therefore became a matter of interest for many engineers, designers and researchers.

2.2   Previous Study for Segmental Bridge Under Static Load Test

As stated by Moses et al. [4] the determination of the level of optimal test loading is important because a high-test load near or above the design live load is proof of strength of the capacity so that possibilities of lower strength could be effectively eliminated based on the results. The several critical load cases were planned and performed in order to find the maximum response such as the stresses, displacements, cracks width and accelerations. If the results are not satisfactory, an initiative and urgency repair or strengthening should be performed to upgrade the structural safety as required by the design level.

The integrity assessment of safety, durability and load carrying capacity of a new segmental precast SBG bridge at The New Haengju Grand Bridge in South Korea which has been reconstructed by incremental launching method after the collapsed during construction had been studied by Hyo-Nam et al. [5]. Thus, proving that, it is important to assess the integrity of the bridges based on established static and dynamic load testing with effective instrumentations for the monitoring of construction control and maintenance before opening to traffic. The proposed procedure for the field load testing and integrity assessment of a precast SBG bridge is shown in Fig. 2.1.

Other than that, Robert et al. [6] conducted a study on several spans of San Antonio "Y" Project, Phase IIC using live load tests before opening to traffic. The bridge consists of mainly precast SBG installed by using span-by-span method and used the combination of internal and external post-tensioning tendons. Span involved were instrumented to measure strains in external post-tensioning tendons, overall span deflection and concrete surface strains during the field investigation. To imitate AASHTO MS18 (HS20-44), loaded 7-yd short-bed dump trucks are positioned on the bridge as the loading. The data of external tendon strains, concrete surface strains

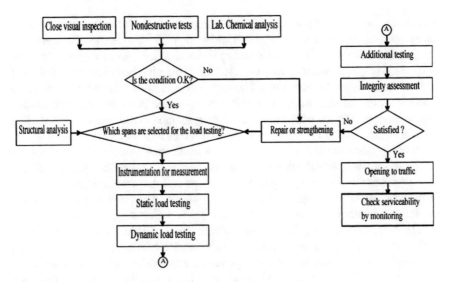

Fig. 2.1 Flow diagram of field testing and integrity assessment [5]

and span deflections were compared with analytical predictions. Under static live load, the bridge behaves very predictably.

Baraka and El-Shazly [7] in their study for the El-Maadia Bridge in Egypt, the Global Positioning System (GPS) is presented as a viable tool to monitor the bridge deformation in Egypt. The static loading test was performed at the completion of the construction of the bridge and precise levelling was the fundamental deformation monitoring technique suggested for the test. The results from GPS were compared to those of precise levelling. Based on the results, it was feasible to continuously monitor selected points along the bridge and also to overcome problems relating to levelling benchmarks. The precise levelling proved to be more effective instead of deflectometers due to deficiencies such as the change of temperature and humidity and gives accurate results for deformation. The results of GPS levels as compared to the results of precise levelling shows that GPS can be used to monitor the bridge deformation during static load test.

Another study was reported by Karoumi et al. [8] on the comprehensive static and dynamic load tests for The New Svinesund Arch Bridge, Sweden—Norway that was performed just before the bridge opening to traffic in June 2005. The bridge is the main route of the European Highway between Gothenburg in Sweden and Oslo in Norway with a total length of 704 m. The static load test was performed by using eight of 25 tonne lorries with known dimensions and axle weights that was positioned according to seven different loading patterns. The lorries were positioned back-to-back in order to have a symmetric loading and for maximum load effect. The important loading patterns were repeated three times in order to verify the reliability of the results. Since traffic load effects are of major interest in this study, many unloaded bridge measurements were made during the day and temperature effects are removed by linear interpolation. The vertical displacement and strain of the bridge was also measured.

Qing Shan Bridge in China as studied by Haleem et al. [9] focused on a prestressed concrete box girder under static load test. The bridge has been operating for more than 29 years since August 1982. The bridge length is 314 m and consists of 13 spans of which the main bridge of variable cross-section of prestressed concrete box girder was constructed with cantilever casting method. The deterioration of Qing Shan Bridge is due to the increase of internal forces that may be a result of higher loading or due to the severe climatic and environmental weather changes that can reduce the cross-sectional resistance to external loads. The static load test is designed to evaluate the load carrying capacity of a bridge and the structural properties such as the concrete strain in order to study the live load effect on longitudinal and vertical deflection of control concrete section by using precise levelling deck method.

The bridge design load level of the vehicle is the tri-axial loading dump trucks with the total weight of approximately 245 kN for each of four trucks, as the standard loading shows the influence line loading consider a horizontal arranging of two lanes, in order to predict the impact and unbalanced live load effect and control section internal forces. The actual load pattern was applied on the bridge twice and the final value of the test point deformation represents the average repeated loading. According to Haleem et al. [9] the principle of equivalent load method, the experimental effects of load and design load effect should be equivalent, and the efficiency of the test load should meet $\eta > 0.85$ (η = calculated value or measured value) considering that the bridge has been used for nearly 29 years. The longitudinal strain, stress, vertical deflection and load carrying capacity of the prestressed concrete box girder under static load test was measured, indicating that the stiffness, the overall deformation, integrity and performance are in good working condition.

The Double-Line Road Bridge, Nanjiang Port of Tianjin Port as reported by Zhao et al. [10] stated that the static load test has been performed in order to analyze the construction quality, safety and structural performance of the bridge that is mainly used for the carriageway of goods. The actual load carrying capacity of the bridge can be determined further. The prestressed concrete box girder bridge with variable cross-section has a total width of 26.5 m. The height of the girder is 4 m at bearing point and 2 m at middle of main-span. The rows of three axle's weight of 75 tonne with 10 m spacing under the design loads were considered in this study. Four control sections to test strains of main-span for the maximum positive and negative moment were located in the middle of piers and on the top of piers. The three load test conditions were calculated and analyzed to examine the bending capacity of the control sections under loading. They also reported that the efficiency factors, η of all working conditions meet the standard requirement of $0.8 \leq \eta \leq 1.05$, according to "Test Method of Long-span Bridge". The value of deflection increment in all test conditions or namely Deflection Analysis (DA) indicates that this bridge structure is in good elastic working state.

2.3 Elastic Behaviour, Displacement, Stress and Strain of Segmental Bridges

A fracture mechanic approach to determine top flange strength proposed by Kaneko and Mihashi [2] is based on cracking and crushing of the concrete. It shows that the strength of a top flange is approximately half of the concrete compressive strength when the normal stress is about half the concrete compressive strength and the shear strength diminishes to zero when the normal stress reaches the concrete compressive strength.

Ahmadi et al. [11] produced a model that combines partial tangential slippage of top flange as a function of the normal displacement at top forty-four flanges. The elastic properties were used to determine tangential displacements when the shear stress is less than the shear strength of a top flange as determined from cohesion and friction properties according to Coulomb's relation.

Turmo et al. [12] studied the formula bending capacity of match-cast dry joints for precast segmental bridges. They discuss a performance of concrete segmental bridge with different depth-to-span ratio and focusing on the displacement behaviour of single unit SBG dry joints under ultimate limit state conditions. Then they compile the varying formulations that were used to evaluate the top flange strength. The formula was adapted to the safety factor provisions set out in Eurocode 2. According to Turmo, the load applied is transmitted across the joint by two qualitatively and quantitatively different mechanisms. The former, which refers to the friction stress arising when two flat compressed surfaces try to move relative to one another, is proportional to the compression forces involved and this proportionality factor is known as friction coefficient.

Kim et al. [13] studied the behaviour of prestressed concrete in static condition that spliced with box segment of precast to compare it with monolithic girder. Displacements, centre of gravity, stress, strain and internal behaviour of spliced girder in static characteristic are presented. It is shown that the monolithic girder in elastic range less compared to the spliced girder. The spliced girder also shows an identical behaviour to monolithic up to 900 kN that is equivalent to service load. The initial load of yield for monolithic girder was 25% more too spliced girder but similar to the post cracking behaviour. A stress of concentration at central segment of the spliced girder shows a result due to no reinforcement across the joint but a typical crack pattern exhibited for spliced girder.

Amanat et al. [14] studied the cause of crack of concrete prestressed web, deck and pier head of the bridge by three-dimensional finite element technique. Effect of temperature causes the crack to widen instead of shrinkage crack during initial stage. A sudden of temperature drop at the bottom of the deck produces a stress that closes the formation of longitudinal crack at centre of the deck. It is reported that there is no significant crack for cantilever side within the top of the depth slab. The deterioration is expected to be slow due to the temperature of the deck.

In general, vertical displacement is one of the parameters that need to be assessed more because their information reflects the overall response and performance of the bridge span in service. Sousa et al. [15] in their studies informed that the vertical displacement can be used to monitor the bridge performance under short-term and long-term observations. The levelling system is a suitable method and one of the most widely used to measure vertical displacements on a river-bed. The aim of this study is to demonstrate the suitability of the polynomial approach to estimate the deflection curve of full-scale concrete bridges. The inclinometers over the supports and strain gauges at mid-span and near the supports were used and these measurements are concerned with short-term observations during load test. Thus, the comprehensive analysis has been carried out in order to evaluate the suitability of polynomial functions as an approximate solution for deflection curves. It was not limited to the effect of boundary conditions, but to the effect of the order of polynomial functions on the results' accuracy. The polynomial function is calculated based on two types of information previously known as; (i) intrinsic characteristics of the bridge's behaviour, namely zero vertical displacement over the supports and null curvature over the outer supports and (ii) curvature and rotations based on measurements performed with strain gauge and inclinometer sensors. Under the simply supported beam, the condition of L/2000 was established as the maximum deflection in order to ensure elastic behaviour during load test.

It is often encountered in practice that the long-term deflections of prestressed bridges are greater than the deflections expected in the design. Navratil and Zich [3] in their studies informed that a substantial number of reasons for excessive deflection arise from technological errors. Particularly the enlargement of water content in concrete mixture, insufficient modulus of elasticity, strength or unit weight resulting from a poor quality of concrete, a wrong sequence or time schedule of construction steps, geometrical imperfections and higher prestressing losses, unsatisfactorily stiffness of temporary supports or poor anchorage, improper and short curing and wrong estimate of the relative ambient humidity. Another reason for unexpected deflection is the omission of some phenomena during structural modelling of a structural element. It concerns for example shear lag and the influence of shear on deformation generally, eccentric position of prestressed and non-prestressed reinforcement, the contribution of non-structural members such as parapets and asphalt wearing surface to the load carrying capacity, stiffness reduction due to cracks development and friction in bearings.

2.4 Previous Finite Element Analysis (FEA) Studies of Segmental Bridges

Antonio and Mari [16] studied the condition of composite bridge by considering different construction process of long-term and short-term deflection, crack, stress, and strain of the concrete slab. The effective of post-tensioned concrete slab to produce

long-term pre-compression was approved to capture the effect of structural performance in the behaviour of composite bridge that lead to the most restrictive design
criteria to a suitable construction solution. The effect of creep and shrinkage of concrete due to distribution of bending moment produced, the negative moment up to
14% over the supports and caused 50% reduced in positive at the centre span. The
stress of slab reinforcement in the centre span also increases 100% from a value of
compression stress and becomes significant tension. Time-dependent crack will also
be produced due to the dead load over the centre of the support that results in creep
and shrinkage due to the increase of bending moment. For a composite of prestressed
bridge, a crack will take place at the support for long-term which the support zone is
cast and post-tensioned. The width of the crack was measured and compared to the
solution of non-prestressed.

Massive end diaphragms are usually used to anchor the external post-tensioning
tendon specifically for the concrete box girder. However, there are several conditions
of distress of the diaphragms that has been reported. Wollmann et al. [17] used the
results from experimental and also analytical on the end diaphragms behaviour when
applied to external tendons anchorage. Physical test of half-scale model and finite
element model are included. The Finite Element Method (FEM) is an effective and
fast method to represent the force flow in the structure used for design. Although
the ultimate strength prediction using finite element are safe, it is quite conservative.
Thus, several trials need to be conducted to achieve the adequate reinforcement
arrangements to control crack.

Rombach [1] examined the behaviour of segmental bridge and the forces in the
joints finite element calculations taking into account the non-linear behaviour due to
the opening of dry joints under tension. A real existing single span segmental bridge
of the elevated highway, Second Stage Expressway System in Bangkok with external
post-tensioning was modelled as shown in Fig. 2.2. This structure is used as data from
a full-scale destructive test with dry joints and external tendons to verify the results of
the complex numerical simulations. The hollow box girder was modelled by 4 nodes
shell elements and the tendons by non-linear truss elements. The opening of the dry
joints is modelled by interface elements while the epoxy glued joints by longitudinal
linear spring elements. This study shows that the indentation of the joints or namely
shear keys can be neglected in numerical calculations if the structure is loaded by
bending only. A good agreement between the numerical results and the test data can
be seen in this study. It can be concluded that the finite element model is capable to
model the real behaviour of a segmental bridge.

Hongseob et al. [18] studied on the crack control during the construction of precast SBG bridges. The theoretical predictions were achieved using the commercial
structural analysis 3-D FEA program, MIDAS which was developed in Korea. The
numerous longitudinal hairline cracks occurred at the bottom slabs of the precast
SBG during the initial service life. A Non-Destructive Test (NDT) method for concrete measurement by using the Ultrasonic Pulses Velocity (UPV) and FEA was
performed in this study. The concrete section and prestressing steel bars were modelled by eight nodes solid element and bar element in order to study the risk of crack
propagation caused by the dead load bending moment. The material properties used

Fig. 2.2 Finite element mesh of a segment

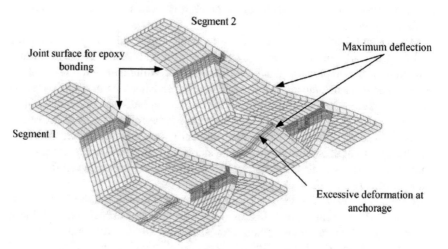

Fig. 2.3 3-D deformation of the segments after temporary prestressing

in this analysis were the existing design data and NDT results. The authors decided to model using the roller condition at the supports for the two segments positioned on a falsework truss while the end section of the previously launched segment nearest to the pier was modelled with hinges. In this analysis, the deformation shapes obtained from the FEA for each segment after external prestressing is shown in Fig. 2.3.

Hongseob et al. [18] also studied a prestressing position of anchors, sequence of prestressing and thickness behaviour of bottom slab to the girder section. An excessive of deformation and tensile stress during external temporary of prestressing due to adjacent bonding of segment box also have been studied. A tensile stress of bottom segment really affects due to the external prestressing of anchorage position and thickness of bottom slab. A sequence to move the position of anchor is proposed in order to decrease the deformation and stress during external prestressed. This proposal is found to be the best technique compared to the others to reduce formation of cracks when considering the cost of construction.

Robertson [19] studied the vertical deflections, span shortening and bending strains for both short-term and long-term behaviour and performance of the prestressed segmental concrete bridge viaduct that measured during a load test eight years after the completion of construction. The measured readings were taken specifically at mid-span and end-span area. Under short-term analysis, the deflections and bending strains that measured during a load test on the viaduct were successfully predicted using a 3-D linear elastic beam element model in SAP 2000, FEA program. Furthermore, the Vibrating Wire Strain Gauges (VWSG), support tiltmeters and optical surveys on the top surface of the viaduct for the baseline system were reliable and accurate for both short-term and long-term monitoring.

Algorafi and Ali [20] studied the effect of torsion on the structural behaviour of Segmental External Prestressed (SEP) concrete beams of the dry joint by experimental and analytical study. A (FEM) study on simply supported bridges using ANSYS finite element software with three different types of elements such as three-dimensional (3-D) cube element, two-dimensional (2-D) interface element and one-dimensional (1-D) link element have been adopted in the analysis. The authors indicated that the most important feature of the FEM is that it has been calibrated with experimental tests which were specifically designed to fail. Further to this analytical model, there are three types of non-linear behaviour such as geometric, material and contact non-linear that was considered. In this study, the responses were investigated in term of deformations (axial and vertical), strain variations, failure load and failure mechanism at different locations. The authors finally concluded that the proposed analytical method can satisfactorily predict the behaviour of SEP concrete box girder beams with or without torsional effect up to the ultimate loading capacities.

Parapets positioned on the deck surface of the bridges are generally known as barriers [21]. Usually the barriers are not included in the structural analysis model for load rating and design purpose. Barriers functions are to withstand the impact due to collision of vehicles, therefore they are not included in the primary structural members. During collision event, barriers will withstand some structural damage and no longer strengthen the bridge deck. Nevertheless, during contact the barriers absorb and distribute the applied load and act as fully functioning structural members. Therefore, the amount of the contribution of the appurtenances, specifically parapets, to the deck strength is an interest to study. Load rating for concrete segmental box girder usually involves elastic limit. For oversized load, it is determined by calculating the transverse analysis of the bridge neglecting the extra strength of fully complete appurtenances. Transverse analysis usually uses Homberg charts to calculate the maximum moment due to the live load. However, this moment estimation lack of consideration such as neglecting the appurtenances, produce conservative load rating for the bridges.

Kuhn [21] developed a 3-D bridge model using LUSAS to produce the qualitative and quantitative values that influence the barrier distribution of live load for three segmental box girders located at Florida Keys. The data obtained from these models are compared to the data obtained from Florida Department of Transportation (FDOT) load test that was conducted on the actual bridges and also to Homberg influence surface prediction. By modifying the current load rating method for structurally sound bridges, including the fully functional appurtenances would interest the authorized agencies and also proves beneficial to the parties that transports oversized loads.

Gupta et al. [22] presented a parametric study for deflections, longitudinal and transverse bending stresses and shear lag for different box girder bridge cross-sections namely rectangular, trapezoidal and circular. Commercially available software SAP 2000 has been used to carry out linear analysis of these box girders. The 3-D elements also have been employed to analyze the complex behaviour of different box girders. The linear analysis was carried out for the dead load (self-weight) and live load for zero eccentricity as well as maximum eccentricity at mid-span area. In this study, it is found that the rectangular section is superior to the other two sections. They proved that the simple beam theory is a crude approximation for analysis of box girder bridge sections such as the longitudinal bending stress distribution in wide flange girders. The mid-span displacement results are compared with the literature results and it can be concluded that the developed model predicts reasonable results.

Zakia has studied the comparison behaviour by using beam and shell element model of straight and curved box girder bridge by examining the stress pattern obtained using static 3-D finite element modelling. The straight box girder transferred stress symmetric longitudinal from one end to another compared to curved box girder which is not symmetric. It shows that the left top flange and web (inside) shows more stress than the right. The stress value can be obtained at every node point for shell model compared to beam model where the 39 stresses are obtained only at the top and bottom. The stress also increased as the spacing of internal cross frame is reduce due to the effect of spacing of bracings.

To determine the influence of truck live load at the Seabreeze Bridge, the precast post-tensioned SBG was instrumented and studied. Maguire et al. [23] used strain gauges that were installed at the interior and exterior of the box girder to determine the transverse response. Then, the result was compared to the analysis using 2-D frame models with a combination of plate surface influence to determine the bending moments in webs, cantilevers and bottom slab. The result shows that experimental values of strain were 90% smaller than the predictions of the analysis. Top slab and cantilever strains were predicted with high accuracy shows more forces spreading at the member than the results from the frame analysis. Near mid-span cantilever strains found to be 60% higher than near a pier. It can be concluded that the traffic fence is stiffer when located near the pier and has greater share with the cantilever live load compared to the mid-span cross-section. Several recommendations were made to increase the analysis accuracy for the simplified frame model such as adding springs at the support to model alteration and on the wing to model the fence as well to increase the longitudinal depth of model near the bottom cross-section.

Gouda [24] conducted a study on the curved bridge single cell box girder. Sample of box girder bridge was chosen from literature as a validation to the finite element modelling method. The sample model of box girder was analyzed in SAP 2000 and the results are very near to the result of the literature. For the parametric study, SAP 2000 was used to model the five box girder bridges. Cross-section, span length and material properties are set as constant in this study. However, radius of curvature was manipulated. Curvature of the bridges differs in horizontal direction only. The models were subjected to dead load (self-weight) and dynamic load of IRC class A tracked vehicles. Static analysis of dynamic load and dead load were conducted. Data of top and bottom longitudinal stress, torsion, bending moment, fundamental frequency and deflection were recorded. The data collected in curved bridge were compared with straight bridge. From the data, parameters such as bending moment, deflection and torsion increases directly the proportional of the curvature of the bridge.

2.5 Previous Transversal Slope Studies of Segmental Bridges

Transversal slope or so-called crossfall gradient analyses have been performed using various software packages on a routine basis in bridge engineering. However, this analysis does not appear officially in many literature reviews. The recent studies have resulted in the collection of extensive orientation problems to which the transversal slopes effect on deflections, stresses and strains can be referenced. Design Manual for Roads and Bridges [25] for surface drainage of wide carriageways has mentioned via clause 2.7 under effect of water on carriageway surfaces that "The UK standard minimum crossfall is 2.5%. This is one of the higher national standards and is intended to achieve efficient removal of water from carriageways, including undulations caused by rutting. Super elevated sections will generally have crossfalls equal or greater than 2.5%, however, areas of low crossfall will occur at super elevation rollovers".

The geometric modelling process of a box girder deck for integrated bridge graphical system was presented by Zita [26]. The possible conclusion by the author is that the deck shape is influenced by two longitudinal geometric components:

1. The morphologic evolution of the deck depth and the thickness of the slabs and webs along its longitudinal axis.
2. The geometry of the layout of the road where the bridge is inserted.

In those systems, the transversal sections of an object developed along a longitudinal axis were always generated on an orthogonal orientation in relation to the longitudinal axis as shown in Fig. 2.4. In the previous studies, the spatial orientation of a cross-section in a deck was always vertical and it cannot be generated by the traditional modelling systems. The generation of 3-D FEM meshing was needed to obtain the correct configuration (interior and exterior) of a box girder deck. From

Fig. 2.4 Cross-section with different shape and transversal slope presented

this study, it can be concluded that the principal advantages of using the developed bridge deck are as follows:

a. A considerable reduction in the time inherently the graphical documentation of the deck that is included in a bridge design.
b. The possibility to actualize the deck geometric database easily at the conceptual design phase such as to minimize deck shapes.
c. The 3-D deck models cross-section will correctly define the shape, orientation and position over the traditional method.
d. The complexity and time taken under 3-D meshes is reduced by the automatic generation of 3-D finite element meshes.

2.6   Summary of Previous Studies

This chapter presents an overall literature review of static load test by experimental, numerical and combination of both studies. These literatures provide information about the performance and behaviour of the precast SBG before and after opening to traffic. From previous studies, it can be seen clearly that many researchers tend to analyze the SBG in one span by considering post-tensioning effect (internal and external) of load. From the analysis, the behaviour of one span of the SBG bridges can be determined. Nevertheless, single segment of the SBG needs to be carried out to distinguish the behaviour of one SBG to another under transverse analysis. By considering the linear analysis, superimposed dead load such as parapets loading and segment's orientation either transversal or longitudinal gradient in this study, the result can be more accurate in determining the single span deflection under short-term analysis. Table 2.1 shows the chronology of previous studies for precast SBG bridges.

Table 2.1 Chronology of previous studies for precast SBG

Researcher	Year	Studies
Muller and Podolny [27]	1982	Construction and design of prestressed concrete segmental bridges
Mathivat [28]	1983	The cantilever construction of prestressed concrete bridges
Moses et al. [4]	1994	Applications of field testing to bridge evaluation
Levintov [29]	1995	Construction equipment for concrete box girder bridges
Hyo-Nam [5]	1998	Field load testing and reliability-based integrity assessment of segmental PC box girder bridges before opening to traffic
Kaneko and Mihashi [2]	1999	Analytical study on the cracking transition of concrete shear key, materials and structures
Antonio and Mari [16]	2000	Long-term behaviour of continuous precast concrete girder bridge model
Wollman et al.	2000	Anchorage of external tendons in end diaphragms
Robert et al. [6]	2001	Live load tests of the San Antonio "Y" Project
Ahmadi et al. [11]	2001	A discrete crack joint model for nonlinear dynamic analysis of concrete arch dam
Rombach [1]	2002	Precast segmental box girder bridges with external prestressing—design and construction
Zita [26]	2003	Geometric modelling of box girder deck for integrated bridge graphical system
Jazlan [2]	2006	Construction of precast segmental box girder bridge using overhead gantry
Algorafi and Ali [20]	2008	Effect of torsion on externally prestressed segmental concrete bridge with shear key
Kumar et al. [30]	2008	Automated geometry control of precast bridges
Kuhn [21]	2008	Transverse analysis and field measurement of segmental box girder bridges
Kim et al. [13]	2008	Overview and applications of precast prestressed concrete adjacent box-beam bridges in South Korea
Ryall [31]	2008	Loads and load distribution studies on bridge structures
Amanat et al. [14]	2010	Cracks in the box girders of Bongobondhu Jamuna Multipurpose Bridge—Identification of causes based on finite element analysis
Gupta et al. [22]	2010	Parametric study on behaviour of box girder bridges using finite element method
Begum [32]	2010	Analysis and behaviour investigations of box girder bridges

(continued)

Table 2.1 (continued)

Researcher	Year	Studies
Dereck [33]	2011	Live load test and finite element analysis of a box girder for the long-term bridge performance program
Haleem et al. [9]	2011	Evaluation behaviour of Qing Shan concrete bridge under static load test
Zhao et al. [10]	2011	Static test analysis of a bridge structure in Civil Engineering
Mosley et al. [34]	2012	Reinforced Concrete Design to Eurocode 2
Gouda [24]	2013	Study on parametric behaviour of single cell box girder under different radius of curvature
Navratil and Zich [3]	2013	Long-term deflections of cantilever segmental bridges
Sousa et al. [15]	2013	Bridge deflection evaluation using strain and rotation measurements
Baraka and El-Shazly [7]	2005	Monitoring bridge deformations during static loading test using GPS
Hongseob et al. [18]	2005	Practical crack control during the construction of precast segmental box girder bridges
Robertson [19]	2005	Prediction of vertical deflections for a long-span prestressed concrete under short-term and long-term analysis
Turmo et al. [12]	2006	Shear strength of match-cast dry joints of precast concrete segmental bridges
Karoumi et al. [8]	2006	Static and dynamic load testing of the New Svinesund arch bridge

References

1. G. Rombach, Precast segmental box girder bridges with external prestressing-design and construction. INSA Rennes, 1–15 (2002)
2. Y. Kaneko, H. Mihashi. (1999). Analytical study on the cracking transition of concrete shear key. Mater. Struct. **32**(217), 196–202
3. J. Navratil, M. Zich, in *Long Term Deflections of Cantilever Segmental Bridges*. Baltic J Road Bridge Eng **8**(3) (2013)
4. F. Moses, J.P. Lebet, R. Bez, Applications of field testing to bridge evaluation. J. Struct. Eng. **120**(6), 1745–1762 (1994)
5. Hyo-Nam et al., *Field Load Testing and Reliability-Based Integrity Assesment of Segmental PC Box Girder Bridge Before Opening to Traffic*. Hanyang University Ansan, South Korea (1998)
6. C. L. Roberts-Wollmann, J. E. Breen, M. E. Kreger, Live load tests of the San Antonio Y. J. Bridge Eng. **6**(6), 556–563 (2001)
7. M.A. Baraka, A.H. El-Shazly, in *Monitoring Bridge Deformations During Static Loading Tests Using GPS*. Proceedings of FIG Working Week 2005, Cairo, 16–21 Apr 2005 (2005)
8. R. Karoumi, A. Andersson, H. Sundquist, in *Static and Dynamic Load Testing of the New Svinesund Arch Bridge*. The International Conference on Bridge Engineering (2006)

9. Haleem et al., *Evaluation Behavior of Qing Shan Concrete Bridge under Static Load Test*. School of Transportation Science and Engineering, Bridge and Tunnel Engineering, Harbin Institute of Technology, Harbin City, China (2011)
10. J. Zhao, T. Liu, Y. Wang, Static test analysis of a bridge structure in civil engineering. Syst. Eng. **1**, 10–15 (2011)
11. M.T. Ahmadi, M. Izadinia, H. Bachmann, A discrete crack joint model for nonlinear dynamic analysis of concrete arch dam. Comput. Struct. **79**(4), 403–420 (2001)
12. J. Turmo, G. Ramos, J.A. Aparicio, Shear strength of match cast dry joints of precast concrete segmental bridges: proposal for Eurocode 2. Materiales de Construcción **56**(282), 45–52 (2006)
13. J.H.J. Kim, W.N. Jin, J.K. Ho, H.K. Jae, B.K. Sung, J.B. Keun, Overview and applications of precast, prestressed concrete adjacent box-beam bridges in South Korea. PCI J **53**(4), 83–107 (2008)
14. K.M. Amanat, A.F.M.S. Amin, T.R. Hossain, A. Kabir, M.A. Rouf, in *Cracks in the Box Girders of Bongobondhu Jamuna Multipurpose Bridge-Identification of Causes Based on FE Analysis*. Proceedings of the IABSE-JSCE Joint Conference on Advances in Bridge Engineering-II (2010), pp. 8–10
15. H. Sousa, F. Cavadas, A. Henriques, J. Figueiras, J. Bento, Bridge deflection evaluation using strain and rotation measurements. Smart Struct. Syst. **11**(4), 365–386 (2013)
16. R. Antonio, M.V. Mari, Long-term behavior of continuous precast concrete girder bridge model. J. Bridge Eng. 22–30 (2000)
17. G.P. Wollmann, J.E. Breen, M.E. Kreger, Anchorage of external tendons in end diaphragms. J. Bridge Eng. **5**(3), 208–215 (2000)
18. Hongseob et al., Practical crack control during the construction of precast segmental box girder bridges. Comput. Struct. **83**, 2584–2593 (2005)
19. I.N. Robertson, Prediction of vertical deflections for a long-span prestressed concrete bridge structure. Eng. Struct. **27**(12), 1820–1827 (2005)
20. M.A. Algorafi, A. Ali, Effect of torsion on externally prestressed segmental concrete bridge with shear key. Am. J. Eng. Appl. Sci. **2**(1) (2009)
21. Kuhn, *Transverse Analysis and Field Measurement of Segmental Box Girder Bridges* (2008), pp 20–29
22. P.K. Gupta, K.K. Singh, A. Mishra, *Parametric study on behaviour of box-girder bridges using finite element method* (2010)
23. M. Maguire, C.D. Moen, C. Roberts-Wollmann, T. Cousins, *Load Test and Transverse Analysis of a Precast Segmental Concrete Box Girder Bridge-Transportation Research Broad (TRB)* (No. 12-3204) (2012)
24. Gouda, *Study on Parametric Behaviour of Single Cell Box Girder under Different Radius of Curvature*. Department of Civil Engineering, National Institute of Technology Rourkela, Odisha-769008, India, May 2013 (2013)
25. Design Manual for Road and Bridges-BD 37/01, Volume 1, Section 3, Part 14, Loads for Highways Bridges
26. Zita, *Geometric Modelling of Box Girder Deck for Integrated Bridge Graphical System*. Department of Civil Engineering, Technical University of Lisbon, Av. Rovisco Pais, Lisbon, Portugal. Automation in Construction **12**(1), 55–66 (2003)
27. J. M. Muller, W. Podolny Jr, Construction and Design of Prestressed Concrete Segmental Bridges (Wiley Series of Practical Construction Guides) (1982)
28. J. Mathivat *The Cantilever Construction of Prestressed Concrete Bridges*. (A Wiley-Interscience Publication, Wiley, New York, NY, 1983)
29. B. Levintov, Construction equipment for concrete box girder bridges. Concr. Int. **17**(2), 43–47 (1995)
30. K. Kumar, K. Varghese, K.S. Nathan, & K. Ananthanarayanan, in *Automated geometry control of precast segmental bridges*. The 25th International Symposium on Automation and Robotics in Construction, vol. 26 (2008)
31. Riyall, *Loads and Load Distribution*. University Surrey. ICE Manual of Bridge Engineering, Institution of Civil Engineers. www.icemanuals.com (2008)

32. Z. Begum, *Analysis and Behaviour Investigations of Box Girder Bridges*. M.Sc., Graduate School of Maryland (2010)
33. Dereck, *Live Load Test and Finite Element Analysis of a Box Girder for the Long-Term Bridge Performance Program*. Utah State University, All Graduate Thesis and Dissertations. Paper 835 (2011)
34. W.H. Mosley, R. Hulse, J.H. Bungey, *Reinforced concrete design: to Eurocode 2*. Macmillan International Higher Education (2012)

Chapter 3
Finite Element Analysis of SBG Subjected to Static Loads

Abstract This chapter describes the experimental and FEM in order to investigate the behaviour of erected precast (SBG) subjected to static load such as deflections, stresses and strains. The procedures and processes in conducting the FEM were acquired from experimental data on a selected span of the bridge. The methodology in this chapter is divided into 4 sections. Section 3.1 presents the material properties of the precast SBG. Section 3.2 describes load application operation and deck deflection monitoring. Section 3.3 explains the strain gauge instrumentation and data recording while Sect. 3.4 explains the parametric study on the models for a variety of transversal slopes using FEA.

3.1 Materials and Properties of Precast SBG

This chapter is focused on the elastic behaviour of erected precast SBG by static load test. The aim of this, is to investigate the comparison of deflections for variety of transversal slope (0, 2.0, 2.5 and 3.0%), stresses and strains under three load cases measurements via experimental and Finite Element Analysis (FEA) studies. Type of material used in this FEA has been taken from previous experimental data on selected span at the Second Penang Bridge Project and has been modelled by using FEA software, LUSAS to run the analysis.

The materials used in the precast segment production are required to comply with the project specification requirements. Design grade of concrete SBG had a specified 28-day compressive strength of 55 N/mm² and were specified as Grade 55 while, the precast parapet is specified as Grade 40 (40 N/mm²). Figures 3.1, 3.2 and Table 3.1 shows a cross sectional view, details of the cross-section at mid-span and dimensions of the bridge superstructure. Where, the material properties like cross-sectional area, moment of inertia, distance from bottom to centroidal axis etc. are shown in Tables 3.2 and 3.3.

The angle of webs with vertical axis can be calculated as shown in Eq. (3.1) [6] below.

F. Mohamed Nazri et al., *Precast Segmental Box Girders*,
SpringerBriefs in Applied Sciences and Technology,
https://doi.org/10.1007/978-3-030-11984-3_3

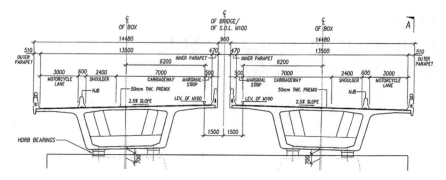

Fig. 3.1 Cross-sectional view of the Second Penang Bridge

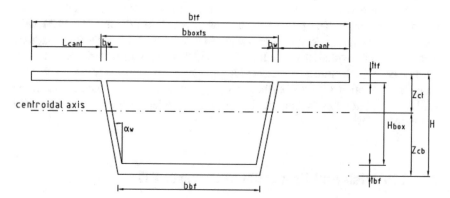

Fig. 3.2 Details of cross-section at mid-span

Table 3.1 Cross-sectional dimensions (refer to Fig. 3.2)

Parameters	Notation	(mm)
Length span	L	55,000
Depth box girder	H	3200
Width of top flange	b_{tf}	14,080
Thickness of top flange	t_{tf}	200
Width web	b_w	450
Width bottom flange	b_{bf}	6200
Thickness bottom flange	t_{bf}	180
Width box top side	b_{boxts}	7800
Cantilever length top flange	L_{cant}	3140
Depth webs	H_{box}	2820

Table 3.2 Values cross-sectional properties of precast SBG

Cross-sectional parameters	Notation	Value
Cross-sectional area of concrete	A_c	7,471,141.69 mm^2
Distance from bottom to centrodal axis	Z_{cb}	1730 mm
Distance from top to centroidal axis	Z_{ct}	1470 mm
Second moment of area of the concrete section	I_c	1.02456E + 14 mm^4
Section modulus bottom	W_b	9,317,832,823 mm^3
Section modulus top	W_t	44,530,190,653 mm^3

Table 3.3 Material properties for concrete, reinforcement and prestressed elements

Concrete properties

1. Concrete Element	Cube strength (N/mm^2)	Maximum aggregate (mm)
Bridge Deck	55	20
Parapets	40	20
2. Unit Weight (All elements)	25 kN/m^3	
3. Coefficient of Thermal Expansion	12E-6/°C	
4. Young's Modulus (E)	35 kN/mm^2	
5. Poisson's Ratio	As BS 5400: Part 4 (Clause 4.3.2)	
6. Creep and Shrinkage (Coefficient & time effects)	As BS 5400: Part 4 (Clause 4.3.2), assume OPC concrete with a cement content of 450 kg/m^3, and a water cement ratio of 0.40	

Reinforcement properties

1. High Yield Steel (Type 2 Deformed to BS 4449 or 4461)	$f_y = 460$ N/mm^2
2. Mild Steel (to BS 4449 or 4461)	$f_y = 250$ N/mm^2

Prestressed properties

*Tendon shall consist of maximum 31 No. seven wire stress-relieved, low relaxation super strand complying with BS 5896
1. Nominal diameter (size) = 15.7 mm
2. Nominal steel area = 150 mm^2
3. Nominal mass = 1.178 kg/m
4. Nominal Ultimate Tensile Strength (Guaranteed UTS, GUTS) = 1860 N/mm^2
5. Guaranteed Nominal Breaking Load = 279 kN
6. Relaxation at 1000 h (70% UTS) = 2.5% at 70% GUTS
7. Jacking force (75% UTS) = 209.25 kN per strand
8. Young's Modulus (E) = 195.0 kN/mm^2
9. Wedge draw-in = 6 mm
10. Coefficient of friction, $\mu = 0.120$ (external tendons); 0.140 (internal tendons)
11. Wobble factor, $k = 0$ radians/m (external tendons); 0.008 radians/m (internal)

Fig. 3.3 3-D view parabolic
path of prestressing strand

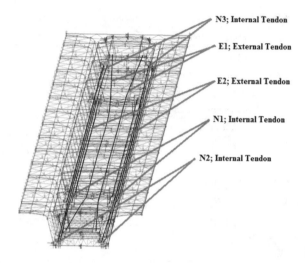

Angle of webs with vertical axis (α_w):

$$\alpha_w = tan^{-1}\left(\frac{(b_{boxts} - b_{bf})/2}{H_{box} + t_{bf}}\right) \tag{3.1}$$

The dead load of the precast SBG per meter is calculated as shown in Eq. (3.2) [6] below. Dead load of the precast SBG per meter (g_{dead}):

$$g_{dead} = A_c \rho_c g \tag{3.2}$$

where

A_c Cross-sectional area (mm^2)
ρ_c Density of concrete (2500 kg/m^3)
g Acceleration due to gravity (9.81 m/s^2).

Mechanical properties such as tensile strength, compressive strength and Modulus of Elasticity are related to the properties and proportions of the constituent materials. Each of the precast SBG was prestressed. The strands used for the post tensioning were low relaxation strands. The strands were jacked to a force of 4540 kN for the initial force, Po (E1; external tendon), 3560 kN (E2; external tendon), 3975 kN (N1, N2 and N3; internal tendon) which were inclusive of friction and stress losses. The prestressing strands followed a parabolic path throughout each span as shown in Fig. 3.3.

3.2 Load Test Execution

Ryall [9] stated that the design should be made on the basis of normal loading (everyday traffic consisting of a mix of cars, vans and trucks) defined as HA loading and check for abnormal loading (consisting of heavy vehicles of 100 tonne or more) defined as HB loading. The HA loading consisted of a Uniformly Distributed Load (UDL) considered together with a single invariable Knife Edge Load (KEL). Abnormal loading shall be 45 units of HB or 30 units of HB in specific combination. BD37/01 [2] says that HB loading requirements derive from the nature of exceptional industrial loads such as electrical transformers, generators, pressure vessels and others likely to use the roads in the area. A requirement for all public roads to be designed for at least 30 units of HB was introduced in 1973 and so-called HB 30. One unit of HB loading is equal to 10 kN per axle.

In this chapter, HB 30 was performed, and the proposed load cases do not reach the most severe live load combination of BD37/01. It is a general consensus among engineers not to test bridge structures to ultimate limitation. The three static load cases shall represent Types HA-UDL, HB 30 and HA-KEL design live load conditions and the attempt is not to test maximum Serviceability Limit State (SLS) capacity of the bridge. The traffic loads considered in the design were HB units with or without adjacent HB units applied to the notional lanes. In accordance with Eurocode 2, different lane live loading arrangements were applied to the two-dimensional (2-D) frame model following the limit state considered as follows:

a. **Live load considered at Ultimate Limit State (ULS)**:

 i.   45 units of HB in combination with another 30 units of HB in another lane or 45 units of HB occupying any transverse position in the carriageway.

b. **Live load considered at SLS**:

 i.   Quasi-permanent combination:
      – No traffic live loads shall be considered under this combination and it is used for long-term effects and appearance of the structure.

 ii. Frequent combination:
      – Typically used for checking decompression and cracking in prestressed member at SLS. 30 units of HB alone in any transverse position in the carriageway were considered.

Different live lane load combinations were considered to derive the worst loading arrangement producing the most severe effect. The worst effects were derived from the load envelope of the different lane loadings. Therefore, the load case under SLS (frequent combination) was preferable. The simulation due to static load cases were carried out by using 50 precast mass concrete block of approximate dimension 1000 mm × 1000 m × 1000 mm that represent HA loading. Where the HB loadings are proposed to be experience span deflection tests of the structural deck to monitor the wind and live load action movements. Prior to the test, every concrete block was

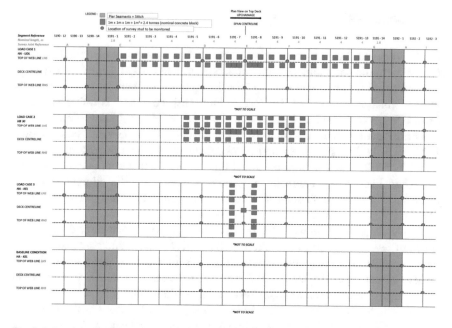

Fig. 3.4 Load application operation and deck deflections monitoring

allocated a number which was printed on the block and weighed individually using an industrial weighbridge, with individual dockets provided. The average weight of the blocks is 2423 kg. The concrete blocks were loaded and unloaded to and from the selected span (South Span S191) deck top slab using a crane on a barge located on the sea adjacent to the south of the span under test. Hence, there were no handlings or traffic loads applied during the loading and unloading operations other than minor personnel loads. The load cases were applied following the sequence as shown in Fig. 3.4. The deck deflections measurements and combination of loadings were taken by three load cases readings under Load Case 1 (LC1), Load Case 2 (LC2), Load Case 3 (LC3) and lastly readings for the final unloaded case which follows the removal of the blocks on completion of the load test in order to confirm the elastic return.

3.3   Strain Gauge Measurement

Strain gauge instrumentation and data recording was carried out following the plan within the proposal see Fig. 3.5 specifically to investigate the response of a single span of the Second Penang Bridge during static loading at mid-span area. Only the strain gauge arrangement at mid-span was taken into consideration. The gauges and data logging equipment were installed with the gauges secured to the concrete surfaces using epoxy. This type of strain gauge that was used in the load test was a "Geocon

Data logging equipment set up at desk

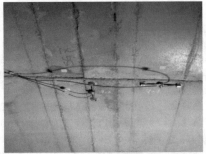
Bottom top slab gauge arrangement at SG7

Bottom slab gauge arrangement at SG5

Web gauge arrangement at SG3

Fig. 3.5 Installed Vibrating Wire Strain Gauges (VWSG) at mid-span area

Model 4000" 150 mm vibrating wire type with a range of 3000 $\mu\mathcal{E}$. Nevertheless, the microstrains ($\mu\mathcal{E}$) results and the associated stresses in the deck at specific locations can be calculated by using the value of Modulus of Elasticity (E) established for the in situ concrete. The result of the measured stress values was tabulated against the theoretical values calculated by FEA and will be discussed in detail in Chap. 4.

3.4 Finite Element Analysis

The FEM is considered to be the most powerful, useful, versatile and flexible method that most of the designers, engineers and practitioners agreed. The precast SBG is a complex structural geometry can be easily modeled using FEM. This simplified method has the capability to deal and to be applied with different material properties, relationships between structural components, boundary conditions, as well as statically or dynamically applied loads.

Line model or so-called Line Beam Method (LBM) was established through a finite element analysis using STAADPro analysis software. The present FEM results by LUSAS were validated by comparing with previous LBM analysis and experimental test results. Viaducts of six spans were modelled with nodes at every segment joint. Member properties of the different elements were input by taking the average properties of the starting and ending joint. The roller supports were modelled at every permanent bearing and one pinned support was put to stabilize the model. The prestress was designed to maintain zero tension across the entire precast joint under SLS.

Gouda [6] in her studies proved that the linear structural response of such bridges can be predicted with good accuracy using FEM to carry out linear analysis of these precast SBG. Three-dimensional (3-D) solid elements were employed to analyze the complex behaviour of different load cases. The linear analysis was carried out for the dead load (self-weight), superimposed dead load (parapets loading) and three live load conditions. All the dimensions such as webs, flanges, diaphragms, top and bottom slabs followed the specification and the approved as-built drawing belonging to Jambatan Kedua Sdn. Bhd. (JKSB). The right method of modelling for this study is important because it will affect the results and also prevent systematic errors from occurring. Therefore, the basic knowledge and information for the modelling must be understood clearly and done in the correct manner to assure the accurate and quality of data. Before conducting a parametric study such as the deflections under a variety of transversal slope, the finite element model was validated using a recent experimental data at selected span.

3.4.1   Description of Finite Element Model

In this chapter, the solid elements by using FEM which was involved in creating models, running analysis and viewing results was created for the Second Penang Bridge. A solid element in the finite element was developed from two-dimensional (2-D) to a specific thickness flange, web and bottom slab in 3-D via LUSAS modelling elements. Most of the solid elements were different in node tetrahedral due to the different type of precast SBG. This design evaluates the effects of permanent loads (parapets loading) and live loads on the precast SBG longitudinally via FEA and theoretical analysis.

Since the box was symmetrical to the centreline, the load arrangement was made in such a way that the adverse load envelope can be obtained for one web only and then made applicable for the full girder. The 3-D solid models were assigned with the appropriate supports to keep the model fully restrained in vertical direction when subjected to eccentric lane load. One of the advantages of solid elements is that it is easy to obtain a better understanding of the behaviour of the structure by using the slice section facility and it helps to produce a detailed design. However, Dereck [3] in his studies informed that one of the disadvantages was that the solid elements can be subjected to condition called shear locking. This shear locking condition occurs when the solid element becomes too stiff under a bending moment and shear deformations occurs instead of bending deformations.

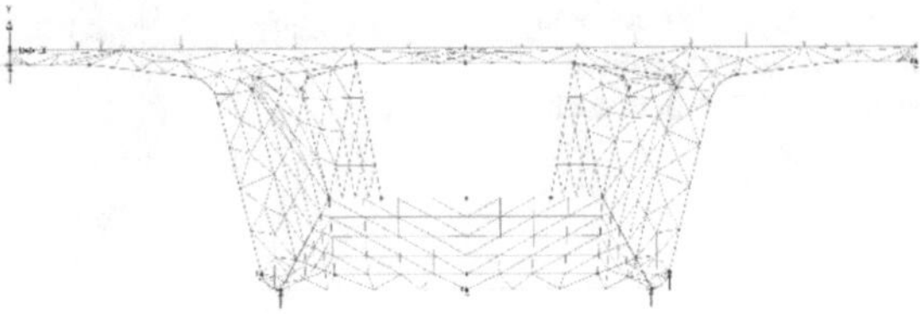

Fig. 3.6 Cross-sectional meshing view of pier segment

Fig. 3.7 Cross-sectional meshing view of typical segment (550 mm web thickness)

Fig. 3.8 Cross-sectional meshing view of typical segment (450 mm web thickness)

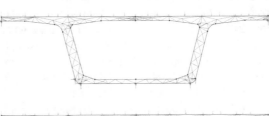

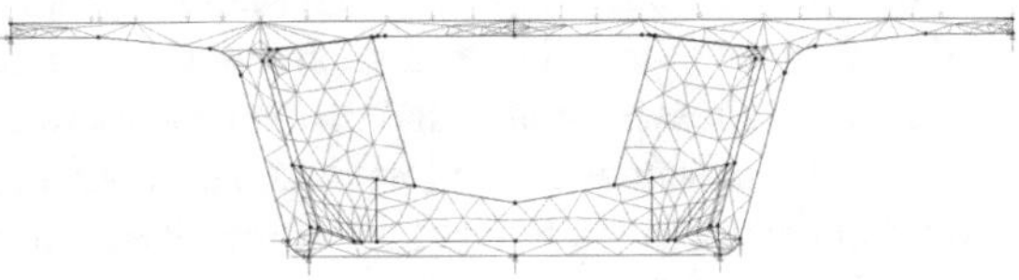

Fig. 3.9 Cross-sectional meshing view of deviator segment

3.4.2 Accuracy of Finite Element Analysis

Generally, several modelling techniques which were considered good practice were implemented such as low aspect ratio, avoidance of small elements and avoidance of small to large element transitions [2]. There was also no skewed geometry and distortional effect considered in these modelling. Figures 3.6, 3.7, 3.8, 3.9, 3.10 and 3.11 shows a cross-sectional view for every single type of precast SBG via the finite element model of the Second Penang Bridge. The right method of modelling for this study is important because it will affect all the results and also to prevent the systematic errors from occurs [7]. Therefore, the basic knowledge and information for the modelling must be understand clearly and done in correct manner to assure the accurate and quality data.

The finite element model was divided into several sections based on type of precast SBG with each section having the same material properties except parapets. These sections included the deck, bottom flange, web girders and diaphragms. Under the pre-processing stage, the 2-D of 200 mm thickness model of typical SBG from AutoCAD 2013 has been transferred to LUSAS and it has been transformed to 3-D

Fig. 3.10 Cross-sectional meshing view of typical segment (350 mm of web thickness)

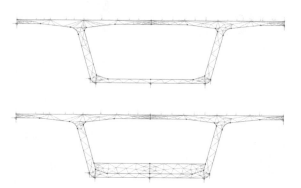

Fig. 3.11 Cross-sectional meshing view of locator segment (350 mm of web thickness)

solid models by using sweep technique form LUSAS. A structure like beam, slab, and column that incorporated time-stage with creep and shrinkage is suitable modelled by LUSAS post-tensioned method [1]. A single unit and multiple tendon of prestress wizard in LUSAS supported for the span by span, balanced cantilever, incremental launching and progressive placement method. Beam, plane stress and solid model of element for active load cases can be calculated by using equivalent nodal load due to tendon prestressing or post-tensioned.

Furthermore, under the post-processing stage, an output part is a program where all the results are viewed after completing the calculation process. The results such as maximum stresses, strains and displacements value can be obtained from output program. These values as well as boundary conditions that changed until a strong correlation between the finite element model and the static load test data was found will be discussed in detail in Chap. 4.

3.4.3   Attributes for Modelling

Models are defined in terms of geometry features which are sub-divided into finite elements for analysis. This process is called meshing and mesh attributes contain information about the element type, element discretization and mesh type. In this study, irregular meshing is used to generate elements on any arbitrary surface. A good initial mesh is obtained by specifying the element size as approximately 1/50 or 2 mm of the diagonal model size. Specifying too small an element size will cause too many elements to be generated and may result in LUSAS using up all the available memory. Specifying too large an element size will cause the meshing algorithm to fail. The success of tetrahedral meshing is dependent on the quality of the surface mesh. If the meshing algorithm fails, set the volume mesh to "From defining geometry" and adjust the element size using 'None' Line and surfaces mesh attributes. If the meshing still fails, try breaking the volume into a number of smaller volumes. These are motivated by convergence, as the mesh is refined and obviously provides unified information on convergence requirements such as completeness,

Table 3.4 Numbers of nodes used for finite element model

Type of precast SBG	Length (mm)	Numbers of segment	Numbers of nodes
Pier segments	2800	2	74
Deviator segments	4000	2	102
Segments with 550 mm thick webs	4000	2	56
Segments with 450 mm thick webs	4000	2	56
Segments with 350 mm thick webs	4000	5	140
Locator segment with 350 mm thick webs	4000	1	31
In situ stitch	1400	2	56
Total			515

compatibility and stability. An important step in finite element modelling is choice of the mesh density. A convergence of results is considered to be achieved when an adequate number of elements are used. This is considered to be achieved when an increase in the mesh density has a negligible effect on the results.

The geometry that was used in this modelling is volume geometry or so-called solid modelling. Volume is the most complicated geometry to analyze in LUSAS compared to line, surface and joint. However, the results are more accurate and precise. The fourteen numbers of precast SBG were modelled using different nodes of solid element due to different types of SBG. Numbers of nodes used in this study are summarized in Table 3.4. In this study, the perfect bonds between solid elements were assumed, so the two elements sometimes shared the same nodes. In this model, the separate models which contained fourteen numbers of segments including of in situ stitch segments (end to end) were combined and came in contact again.

After completing the geometry model, meshing was assigned to the geometry by using line and volume mesh for tendon profile design and SBGs design. Mesh describes the element type and discretization on the geometry. Irregular mesh was used in volume mesh to comply the span to depth ratio in the model while tetrahedral element shape and linear interpolation order were also used in the volume mesh. The materials that were used for the precast SBG elements are concrete grade short-term C50 from BS 5400 while stainless steel ungraded for tendon elements. The concrete grade is in compliance with the strength of the concrete grade that was used for actual casting in segmental bridge construction.

The tendon profile is defined as a spline. Note that sufficient points must be used in the line definition to accurately represent the tendon profile as a series of straight lines or curve lines. Further to that, if the tendon (external or internal) is to be included in the analysis model, thick beam elements are assigned to the lines defining the tendon profile and the concrete beam surrounding the tendon is modelled with plane

Fig. 3.12 Finite element
model representation of LC1

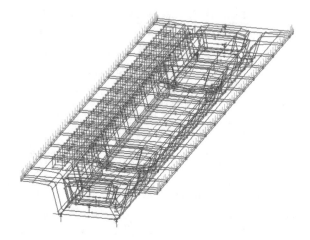

stress or 3-D solid elements. By using this simplified force approach, the effects of prestressed can be defined in a separate linear load case that can be combined with the other load cases to compute the overall structural behaviour.

For the support, fixed support is applied at three nodes at both bottom end of the box girder. This is to make the box girder withstand the applied load and to see the deflection result from the box girder after loading is applied. Then the last attribute that need to be assigned is the structural loading. In this study, global distributed load [5] is used to apply at the top surface of the deck as follows:

1. HA-UDL (LC1), simulated by 50 number blocks placed over the top of one web as one lane over the full length of the span, weighing approximately of 1200 kN (see Fig. 3.12).
2. HB 30 (LC2), simulated by 50 number blocks placed over the top of one web as two lanes over the central section of the span, weighing approximately of 1200 kN (see Fig. 3.13).
3. HA-KEL (LC3), simulated by 17 number blocks placed over the two central segments, weighing approximately of 408kN (see Fig. 3.14).

3.4.4   Comparison with Deflection

Mosley et al. [8] mentioned that in majority cases, particularly where the member is designed to be uncracked under full load, a simple linear elastic analysis based on the gross concrete section will be sufficient to give a reasonable and realistic estimate of deflections. Comparing the FEM deflections with the measured bridge deflections was important because it defined the global behaviour of the bridge. The value of maximum deflection is then compared with simple beam theory value of self-weight. At transfer of displacement, a value of a displacement under dead load and prestress

Fig. 3.13 Finite element
model representation of LC2

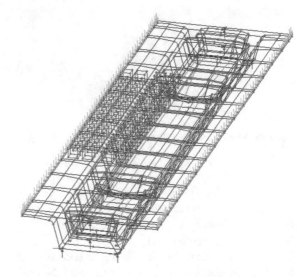

Fig. 3.14 Finite element
model representation of LC3

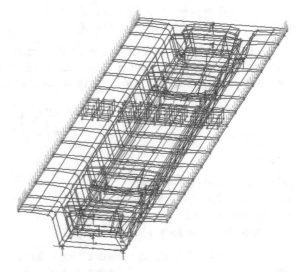

force of tendon either pre-tensioned or post-tensioned method due to initial and final
stress force can be divided under dead load and under prestress load as mentioned
by Mosley et al. [8]. The mid-span deflection due to a uniformly distributed load, w
over a span, L at transfer and application of finishes is determined with the formula
as defined in Eqs. (3.3) and (3.4).

$$Y_a = \frac{5}{384}\left(\frac{W_{\min}L^4}{E_{cm}I_c}\right) - \frac{5}{48}\left(\frac{P_o e}{E_{cm}I_c}\right)L^2 \tag{3.3}$$

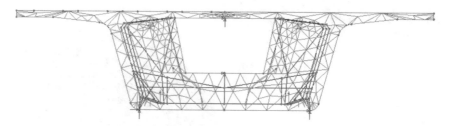

Fig. 3.15 Meshed model of precast SBG with 0% transversal slope

where

Y_a	Deflection at transfer
W_{min}	Self-weight of precast SBG (kN/m)
L	Span length (mm)
I_c	Moment of inertia of concrete section (mm^4)
$E = E_{cm}$	Modulus of elasticity of concrete (kN/mm^2 or Gpa)
P_o	Initial prestress force at transfer (kN)
E	Eccentricity at mid-span (mm).

At application of finishes, assume that only a small proportion of prestress losses have occurred. Therefore, a combination value of a displacement due to transfer stage and load cases applied on top of one span of structure is shown in Eq. (3.4).

$$Y_b = Y_a - \left(\frac{5WL^4}{384EI} \right) \qquad (3.4)$$

where

W	Type of load cases (kN/m)
L	Affected length due to load case (mm)
I	Second moment of area (mm^3)
E	Modulus of elasticity (kN/mm^2 or Gpa)

Please be reminded that an upward deflection has a negative sign and a downward deflection has a positive sign. In order to ensure that the FEA results are in good agreement with the experimental data and theoretical calculation, both meshed model of precast SBG with a variety of transversal slope or so-called crossfall gradient of 2.5% as per actual condition on site, a constant crossfall was studied. Figures 3.15, 3.16, 3.17 3.18 shows a cross-sectional view of precast SBG as mentioned above.

In order to investigate the comparison between the experimental results vs. the FEM, Eq. (3.5) presents the percentage of error to be obtained for the maximum deflections at mid-span area.

$$\%\text{Error} = \frac{\text{FEM} - \text{Experiment}}{\text{Experiment}} \times 100 \qquad (3.5)$$

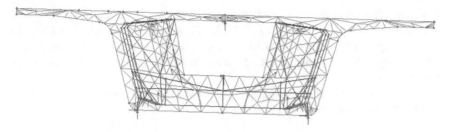

Fig. 3.16 Meshed model of precast SBG with 2.0% transversal slope

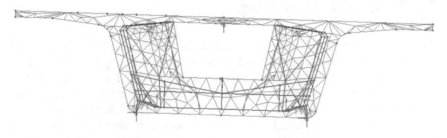

Fig. 3.17 Meshed model of precast SBG with 2.5% transversal slope

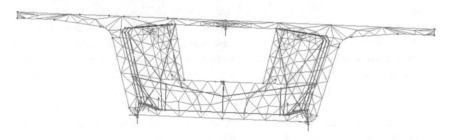

Fig. 3.18 Meshed model of precast SBG with 3.0% transversal slope

3.4.5 *Comparison with Strain*

Based on previous studies, prior to using the finite element model to predict future behaviour, it must be calibrated by comparing its deflections, strains and rotations with the measured field results. From observations of the static load testing on a single span of precast SBG on the Second Penang Bridge Project, the deflections and bending stresses at the mid-span are the main concern to be mentioned in this chapter. The value of Modulus of Elasticity (E) in the structural analysis for the determination of design load effects was the short-term modulus with the exception of differential settlement effects where the long-term modulus was used.

Strain correlation was important, but the strains could have been subjected to local strain behaviour and not necessarily the global response of the bridge. The change in strain values from the FEM prediction was calculated by using the stress from a solid element at a particular load case (LC1, LC2 and LC3). The actual Modulus of Elasticity (E) for the concrete was determined by taking two 100 mm diameter cores from the top slab of the selected span prior to test and the strain was calculated using Hooke's Law.

A comparison of the measured and predicted strains that was calculated from FEM will be discussed in detail in Chap. 4.

3.5 Summary of Methodology

The behaviour of erected precast (SBG) such as deflections, stresses and strains were investigated and observed by experimental and FEA method subjected to static load test. Therefore, the details bridge description, cross-sectional dimensions, materials and properties are the vital part to takes into consideration in this chapter. The elastic behaviour of precast SBG and the comparison of deflections, bending stresses and strains by FEM were validated by experimental work under short-term analysis. In general, vertical displacements are one of the parameters that more often need to be assessed because their information reflects the overall response of the segmental bridge span. However, Vibrating Wire Strain Gauges (VWSGs) are easier to install, but their measurements provide no more than indirect information regarding the bridge deflection. Furthermore, the static load cases (normal and abnormal loadings) and the precise levelling method proved to be more effective to generate span deformation tests for the structural performance of the deck structure. Then, the parametric study on the models for a variety of transversal slopes (0, 2.0, 2.5 and 3.0%) was intended to achieve efficient removal of water from carriageway which take into account the superimposed dead load such as parapets in the design compared with analytical predictions using LUSAS software was studied.

References

1. K.M. Amanat, A.F.M.S. Amin, T.R. Hossain, A. Kabir, M.A. Rouf, in *Cracks in the Box Girders of Bongobondhu Jamuna Multipurpose Bridge-Identification of Causes Based on FE Analysis*. Proceedings of the IABSE-JSCE Joint Conference on Advances in Bridge Engineering-II (2010), pp. 8–10
2. BSI. (2004a). BS EN 1992-1. Eurocode 2, *Design of concrete structures Part 1-1–General rules and rules for buildings (including NA)*. London: British Standards Institution.
3. Dereck, *Live Load Test and Finite Element Analysis of a Box Girder for the Long-Term Bridge Performance Program*. Utah State University, All Graduate Thesis and Dissertations, Paper 835, (2011)
4. Design Manual for Road and Bridges-BD 37/01-Volume 1, Section 3, Part 14, Load for Highway Bridges
5. FEA Ltd, LUSAS version 14. *Modeller User Manual- Engineering finite element analysis and design software,* United Kingdom

6. Gouda, *Study on Parametric Behaviour of Single Cell Box Girder under Different Radius of Curvature*. Department of Civil Engineering, National Institute of Technology Rourkela, Odisha-769008, India, May 2013 (2013)

7. P.K. Gupta, K.K. Singh, A. Mishra, *Parametric Study on Behaviour of Box-Girder Bridges Using Finite Element Method* (2010)

8. W.H. Mosley, R. Hulse, J.H. Bungey, *Reinforced concrete design: to Eurocode 2*. Macmillan International Higher Education (2012)

9. Riyall, *Loads and Load Distribution. University Surrey*. ICE Manual of Bridge Engineering, Institution of Civil Engineers. www.icemanuals.com, (2008)

Chapter 4
Validation of Experimental and Analytical Study Work

Abstract In this chapter, the elastic behaviour and performance of precast segmental box girder (SBG) subjected to static load test is discussed based on the experimental results and finite element method (FEM). By using finite element analysis (FEA), stress and strain could be plotted in (a) graphs to produce stress-strain relationship. Based on the deflection formula, result from the FEA was compared with the calculation and the percentage of error was determined. Besides that, the comparison between the deflection behaviour of precast SBG with a variety of transversal slope or (so-called) crossfall gradient is also discussed in this chapter. A single span of precast SBG bridge model with constant span length, depth, thickness of top and bottom flange, width of top and bottom flange, thickness of web and few variables are tested for comparison of results. The variables are (a): Transversal slope (0, 2.0, 2.5 and 3.0%), and (b) Three types of load cases at HA-UDL (Load Case 1), HB 30 (Load Case 2) and HA-KEL (Load Case 3).

4.1 Elastic Behaviour of Precast SBG

Baseline surveys were taken at 9 am, 12 pm, 3 pm and 6 pm in order to allow for the deck to response to temperature effects during the day. This is done so that deflections under three load cases could be determined by comparing readings taken at the specified load cases time similar to those taken for the baseline readings. From the Unloaded Case 1 and Unloaded Case 2 readings, the differences in time are minimal based on the change of deck elevations recorded.

Measuring the microstrains at the mid-span area that consist of three related strain gauges (vibrating wire type) with different orientations. Vibrating Wire Strain Gauges (VWSG) were installed on the web (VWSG3), top and bottom deck (VWSG5 and VWSG7) at inner side of the bridge at location 3, 5 and 7. The orientation of the VWSG's at each location is shown in Figs. 4.1 and 4.2. Table 4.1 shows the orientation and arrangement of VWSGs that were located in specific locations at mid-span area.

F. Mohamed Nazri et al., *Precast Segmental Box Girders*,
SpringerBriefs in Applied Sciences and Technology,
https://doi.org/10.1007/978-3-030-11984-3_4

49

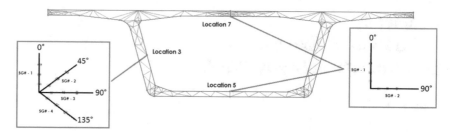

Fig. 4.1 Orientation of VWSGs at location 3, 5 and 7 inside SBG

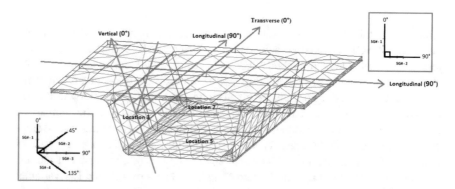

Fig. 4.2 Orientation of VWSGs at location 3, 5 and 7 in 3-D meshing view of typical segment

Table 4.1 Orientation and arrangement of VWSG's at location 3, 5 and 7

VWSG	Location	Installed gauge			
		Longitudinal	Vertical	Transverse	Shear
1	Web	N/A			
2	Web	N/A			
3	Web	3-3	3-1	N/A	N/A
4	Bottom deck	N/A			
5	Bottom deck	5-2	N/A	5-1	N/A
6	Top deck	N/A			
7	Top deck	7-2	N/A	7-1	N/A

*N/A = Not available in this study
*SG = Strain Gauge i.e. SG3-1—Strain gauge number 3 at web in vertical direction
*T = Temperature i.e. T3-1—Temperature effect at web in vertical direction

The bridge was monitored for 24 h in the Unloaded Case 1 as a datum in order to determine the effects of thermal-induced microstrain. The effects of temperature on a precast segmental bridge superstructure are similar to the temperature effects on any bridge superstructure in the longitudinal direction. For consideration of longitudinal temperature differential effects on a simply supported precast SBG in the Second Penang Bridge Project, Figs. 4.3, 4.4, 4.5, 4.6, 4.7 and 4.8 shows a strain and temperature against time at mid-span area where the web, top and bottom deck temperature was increased with respect to the time and strain of the box sections. After completion of the three load cases, all concrete blocks were removed, and the bridge was monitored in the Unloaded Case 2 for a further 24 h. It can be concluded that the return of the values to original values after the static load test confirms the elastic behaviour of the deck. The strain gauge data output shows a cyclic response of the structure to thermal variations during the day and night.

4.2 Comparison with Deflection

By calculating the deflections using the time-related or associated temperature-related baseline survey observations by precise levelling, the values are normalized to be independent of temperature. At the early stage analysis, the FEM confirmed that the parapets loading have a significant impact on the load response distribution so that the correct finite element model can be matched to the segmental bridge data. The deflections were recorded and monitored along the longitudinal direction of the single span precast SBG via FEM. Comparing the FEM deflections with the measured bridge deflections was important because it defined the global behaviour of the bridge.

The theoretical expected deflection data was provided by line beam model (LBM) using STAADPro analysis software. The theoretical manual calculation was also provided as defined in Eqs. (3.1)–(3.5). These equations were used to compare with the actual precise levelling measured data and FEM in order to determine the responses of the deck to the three load cases; Load Case 1 (LC1), Load Case 2 (LC2) and Load Case 3 (LC3) as shown in Figs. 4.9, 4.10 and 4.11. Tables 4.2, 4.3 and 4.4 shows that percentage error in the deflection values obtained from LC1 (2.17%), LC2 (−14.51%) and LC3 (10.28%) are very minimal and conservative. Only 2.5% of transversal slope was considered as per actual span erected on site. The geometric components of the precast SBG such as horizontal alignment (z-axis), vertical alignment (y-axis) and transversal alignment (x-axis) were also considered in this study. The FEM displacement data were close to the experimental data as compared with 0% of transversal slope. But most of the designers and engineers preferred to use traditional method such as LBM due to the small percentage errors. Hence, the finite element model can be considered as validated under the short-term analysis. Several figures of the comparison between the FEM and measured displacements from the static load test can be seen in Figs. 4.12, 4.13 and 4.14. From these figures, it was evident that an excellent correlation existed between the deflection values for the

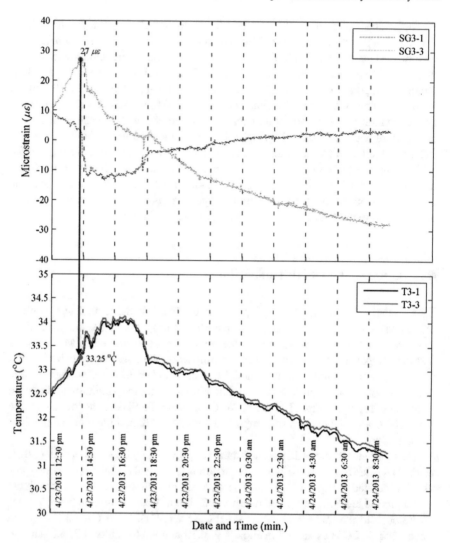

Fig. 4.3 Strain (μ𝓔) and temperature (°C) versus time at mid-span area (VWSG3) under Unloaded Case 1

FEM and the static load test experimental data. It was also proven that when a member is assumed and designed to be uncracked under full load, a simple linear elastic analysis based on gross cross-section is sufficient to give reasonable and realistic estimate of deflections.

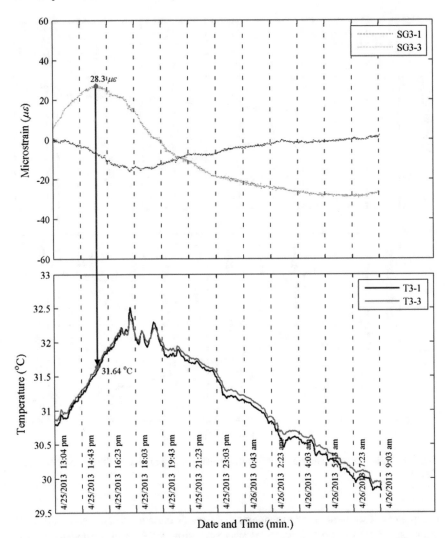

Fig. 4.4 Strain (με) and temperature (°C) versus time at mid-span area (VWSG3) under Unloaded Case 2

The maximum allowable long-term differential settlement between two successive piers was 15 mm as mentioned by design engineer. In LBM analysis, it was assumed that 50% of settlement happened during construction period by erecting segments on top of piers, which was 7.5 mm in differential settlement. Under short-term analysis as obtained in this study, it is proven that the maximum deflections under three load cases were below 7.5 mm. The concrete creep and shrinkage were not considered in this analysis. On the other hand, the support boundary conditions with fixed-end structural support was chosen and assumed that the degree of freedom was

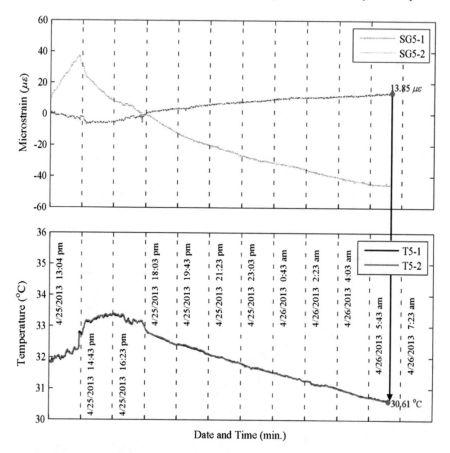

Fig. 4.5 Strain ($\mu\mathcal{E}$) and temperature (°C) versus time at mid-span area (VWSG5) under Unloaded Case 1

completely restrained from movement. The result concluded that the actual deck response under the three load cases was very similar and had a good agreement results to the respective theoretical expected deflections and FEM under short-term analysis.

During the static load test, a linear elastic behaviour is expected and therefore the bridge deflection might be accurately estimated with simple theoretical calculations. Considering the Double Integration Method (DIM) as informed by Mosley et al. [2], the deflection curve of a uniformly distributed loaded is sufficient to give reasonable and realistic estimate of deflection. A single span of precast SBG was considered with constant moment of inertia, geometric and material properties, uniformly loaded and subjected to end forces and moments. Furthermore, the FEM analysis was valid and rational with a smooth curve as shown in Figs. 4.12, 4.13 and 4.14, respectively. Nevertheless, for LC2, the bridge deflection was considerably low compared with the experimental result, but a good conformity between the theoretical calculation,

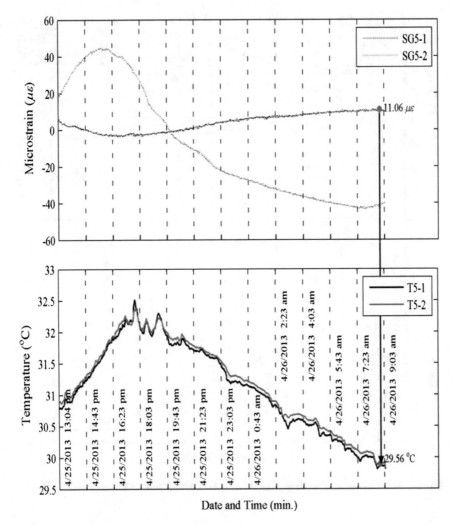

Fig. 4.6 Strain ($\mu\varepsilon$) and temperature (°C) versus time at mid-span area (VWSG5) under Unloaded Case 2

LBM and FEM with different transversal slopes was achieved for that load case, with a maximum percentage error of -14.51% (2.5% transversal slope) and -13.2% (0% transversal slope) respectively. From the result of LUSAS Modeler analysis, it shows that the maximum displacement occurred at the mid-span area of SBG. The red colour shows where the critical displacement was occurred. It can be stated that the middle part of the one span SBG was a critical part where the maximum displacement occurs. Furthermore, this situation was equivalent to the case of precast SBG during the real construction. The most critical component of precast SBG is in the middle part. The FEM analysis proves that the modelling considered the boundary constraints,

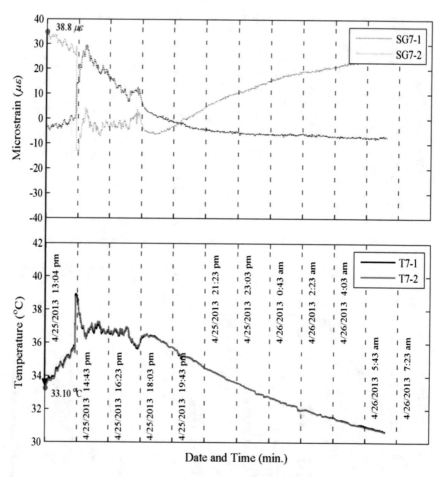

Fig. 4.7 Strain (μℰ) and temperature (°C) versus time at mid-span area (VWSG7) under Unloaded Case 1

namely null vertical displacements over the supports and null curvatures over the outer supports while the experimental curvatures was derived from precise levelling readings. Thus, the deflection of a precast SBG span is highly influenced by the behaviour of cross-sections near the mid-span and support zones.

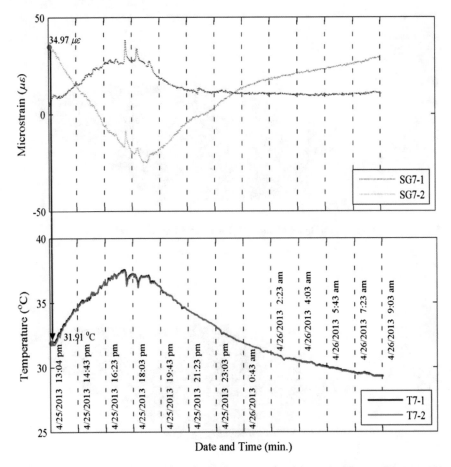

Fig. 4.8 Strain (με) and temperature (°C) versus time at mid-span area (VWSG7) under Unloaded Case 2

4.3 Parametric Study for Different Transversal Slope Under Maximum Deflection Along the Span Length of the Precast SBG

The deflection parameter was recorded for all the bridge models with different transversal slope along the span length of the precast SBG. The horizontal alignment of the model was assumed straight as per actual condition on site. The boundary condition was simply supported and had a single span of 55 m. The geometric and material property was same even for the different conditions.

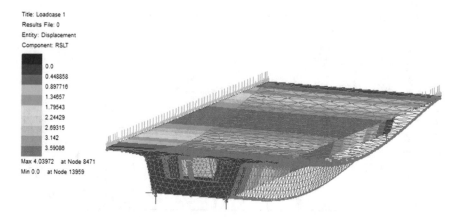

Fig. 4.9 Deflection analysis by FEM under LC1 with 2.5% transversal slope

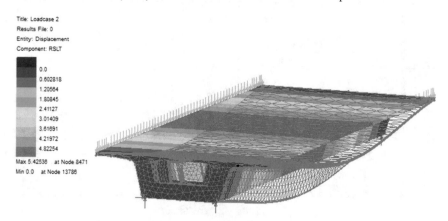

Fig. 4.10 Deflection analysis by FEM under LC2 with 2.5% transversal slope

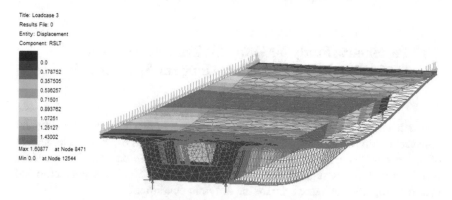

Fig. 4.11 Deflection analysis by FEM under LC2 with 2.5% transversal slope

FEM accuracy was tested under different load cases specifically at mid-span and the contour lines were checked for symmetry. The element x-axis may be rotated by an angle theta, θ (in degrees, °). The maximum percentage error obtained for the displacements under short-term analysis for 2.5% transversal slope was 10.28% in the LC3 which is conservative while the rest of the errors were less than 10% compared to those obtained from the finite element model for mid-span area as shown in Table 4.5.

The experimental results were assumed as the baseline and the maximum percentage error obtained for the displacements was 12.27% for LC3, −13.17% for LC2 and 10.36% for LC1. It can be seen that the deflections obtained using FEM linear analysis are most preferable compared to the deflections obtained using other methods. In order to seek the best fitting for the precast SBG deflection, a parametric analysis was performed along the span length at different transversal slope 0, 2.0, 2.5 and 3.0% in order to prove that the deck displacement correlated to the transversal slope in the actual load test. Thus, it should be noted that the different transversal slope would affect the vertical displacement results under long-term deflections.

From the calculated percentage error, it ranges from −14.51 to 10.28% under 2.5% transversal slope. The ranges are almost less than 10% that was still considered as acceptable [1].

4.3.1   Stress and Strain Analysis at Mid-Span for 2.5% Transversal Slope

Stress and strain develop from the action of the loading applied to the segment due to distortion of the members. A graphical comparison under 2.5% transversal slope was studied in term of vertical deflection, Vertical Bending Stress (VB$_{stress}$), Longitudinal Bending Stress (LB$_{stress}$) and Transverse Bending Stress (TB$_{stress}$) along the span across the cross-section in top, bottom and web sections at mid-span area for live load effects. Due to the limitation of space only the results for Load Case 1 (9 am) are presented here. However, in order to get a more comprehensive insight into the precast segmental bridges stress and strain, this analysis is also supported by results obtained from FEM. Therefore, the estimated bending stresses and strains can be confronted, not only in the cross-sections where the measurement were taken, but also throughout the bridge span length taking advantage of the results from the FEM. The bending stresses (σ) versus time curves that are presented in Figs. 4.15, 4.16, 4.17, 4.18, 4.19, 4.20, 4.21, 4.22, 4.23 and 4.24 shows variation of VB$_{stress}$, LB$_{stress}$ and TB$_{stress}$ across the cross-section in top deck, bottom deck and in the web by using experimental method and FEM. TB$_{stress}$ at web section was not available due to the reason that the strain gauges arrangement monitored longitudinal and vertical stress only (Table 4.1).

Table 4.2 Comparison of deflections obtained in experimental results, LBM, theoretical calculation and FEM under LC1 (1200 kN) (unit: mm)

Survey point	Experimental (Measured on site)	LBM (STAADPro)	Theoretical calculation (δmax) (Eq. 3.4)	0% Transversal slope	Error (%)	2.5% Transversal slope	Error (%)
A	−0.2925	1.00		0	−100	0	−100
B	−0.3050	0.50		0	−100	0	−100
C	−1.3938	−1.00		−0.45054	−67.67	−0.44886	−67.79
D	−3.6813	−4.00		−3.60435	−2.09	−3.59086	−2.46
E	−3.9538	−4.00	−3.4890	−4.08694	3.37	−4.03972	2.17
F	−3.5213	−4.00		−3.60435	2.36	−3.59086	1.98
G	−1.0188	−1.00		−0.45054	−55.77	−0.44886	−55.94
H	−0.0413	0.50		0	−100	0	−100
I	0.3200	1.00		0	−100	0	−100

Table 4.3 Comparison of deflections obtained in experimental results, LBM, theoretical calculation and FEM under LC2 (1200 kN) (unit: mm)

Survey point	Experimental (Measured on site)	LBM (STAADPro)	Theoretical calculation (δmax) (Eq. 3.4)	0% Transversal slope	Error (%)	2.5% Transversal slope	Error (%)
A	0.5425	1.00		0	−100	0	−100
B	−0.1826	0.50		0	−100	0	−100
C	−1.5500	−1.00		−0.61206	−60.51	−0.60282	−61.11
D	−5.6125	−5.00		−4.89648	−12.76	−4.82254	−14.08
E	−6.3463	−5.50	−5.6570	−5.50853	−13.20	−5.42536	−14.51
F	−5.3975	−5.00		−4.89648	−9.28	−4.82254	−10.65
G	−1.2288	−1.00		−0.61206	−50.19	−0.60282	−50.94
H	−0.1200	0.50		0	−100	0	−100
I	0.5338	1.00		0	−100	0	−100

Table 4.4 Comparison of deflections obtained in experimental results, LBM and FEM under LC3 (408 kN) (unit: mm)

Survey point	Experimental (Measured on site)	LBM (STAADPro)	Theoretical calculation (δmax) (Eq. 3.4)	0% Transversal slope	Error (%)	2.5% Transversal slope	Error (%)
A	0.0450	0.50		0	−100	0	−100
B	−0.1400	0.00		0	−100	0	−100
C	−0.3588	0.00		−0.18119	−49.49	−0.17875	−50.17
D	−1.4938	−1.50		−1.44955	−2.96	−1.43002	−4.27
E	−1.4588	−2.00	−1.6860	−1.63075	11.79	−1.60877	10.28
F	−1.2613	−1.50		−1.44955	14.93	−1.43002	13.38
G	−0.1863	0.00		−0.18119	−2.71	−0.17875	−4.03
H	0.0863	0.00		0	−100	0	−100
I	0.3925	0.50		0	−100	0	−100

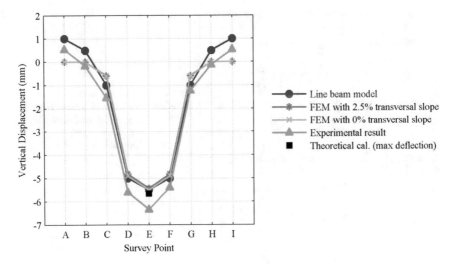

Fig. 4.12 Deflection of the precast SBG under LC1

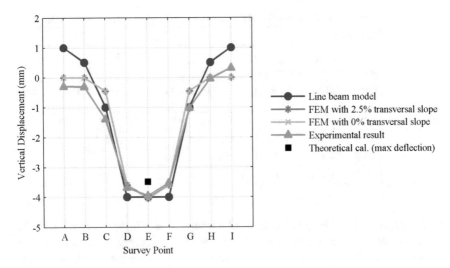

Fig. 4.13 Deflection of the precast SBG under LC2

After seeing Figs. 4.15, 4.16, 4.17, 4.18, 4.19, 4.20, 4.21, 4.22, 4.23 and 4.24 and from the calculated bending stress distribution in case of dead load and live load effects during static load test, it can be concluded that LB_{stress} in bottom deck at mid-span area is the highest value of 2.33 N/mm² followed by LB_{stress} in web with − 0.81 N/mm² and −1.24 N/mm² in top deck. TB_{stress} in top deck is the highest value at 0.99 N/mm², followed by the bottom deck with 0.54 N/mm² while, VB_{stress} in web was 0.45 N/mm².

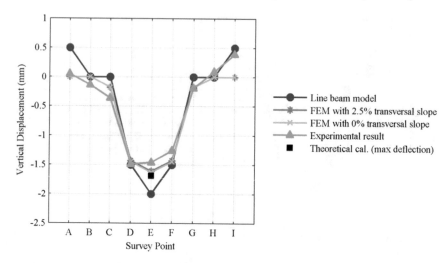

Fig. 4.14 Deflection of the precast SBG under LC3

Table 4.5 Maximum deflection at different transversal slope-FEM and experimental tests

Transversal slope (%)	Load cases								
	LC1			LC2			LC3		
	Maximum displacement (mm)								
	Exp.	FEM	Error (%)	Exp.	FEM	Error (%)	Exp.	FEM	Error (%)
0	−3.954	−4.087	3.36	−6.346	−5.509	−13.20	−1.459	−1.631	11.79
2.0	−3.954	−4.363	10.36	−6.346	−5.511	−13.17	−1.459	−1.638	12.27
2.5	−3.954	−4.040	2.17	−6.346	−5.425	−14.51	−1.459	−1.609	10.28
3.0	−3.954	−4.100	3.71	−6.346	−5.478	−13.68	−1.459	−1.627	11.54

As mentioned earlier, the deflection of a bridge span is highly influenced by the behaviour of cross-sections at the mid-span and near the support zones. Moreover, a failure scenario normally occurs in these zones due to the high strain level. The strains were measured only at the mid-span cross-sections. The bending strains (ε) versus time curves that are presented in Figs. 4.25, 4.26, 4.27, 4.28, 4.29, 4.30, 4.31, 4.32, 4.33 and 4.34 shows variation of Longitudinal Bending Strain (LB$_{strain}$), Transverse Bending Strain (TB$_{strain}$) and Vertical Bending Strain (VB$_{strain}$) across the cross-section in top deck, bottom deck and in the web by using experimental and FEM. The strain gauges data output shows a cyclic response of the structure to thermal variations during testing. The data was recorded at a rate of 1 sample per minute per channel and stored on the data logger.

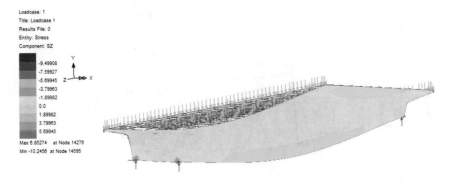

Fig. 4.15 LB$_{stress}$ under LC1

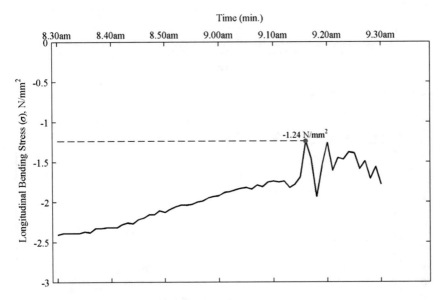

Fig. 4.16 LB$_{stress}$ at top deck (mid-span)

After seeing Figs. 4.25, 4.26, 4.27, 4.28, 4.29, 4.30, 4.31, 4.32, 4.33 and 4.34 and from the calculated bending strain distribution in case of dead load and live load effects during static load test, it can be noticed that, in the bottom deck the LB$_{srain}$ of the mid-span area was the peak value of 42.64 $\mu\mathcal{E}$, then followed by the web LB$_{strain}$ with -14.76 $\mu\mathcal{E}$, where the top deck was the minimum of -22.63 $\mu\mathcal{E}$. For the top deck it was maximum value of 18.04 $\mu\mathcal{E}$m, which is followed by the bottom deck with 9.82 $\mu\mathcal{E}$, while the web was 8.21 $\mu\mathcal{E}$ in the VBstrain. Based on the measurement deck

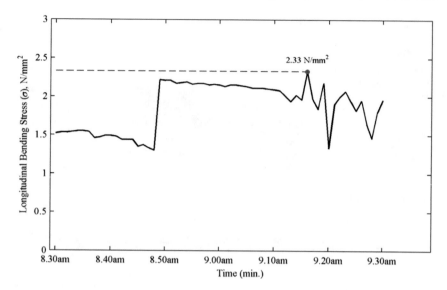

Fig. 4.17 LB$_{stress}$ at bottom deck (mid-span)

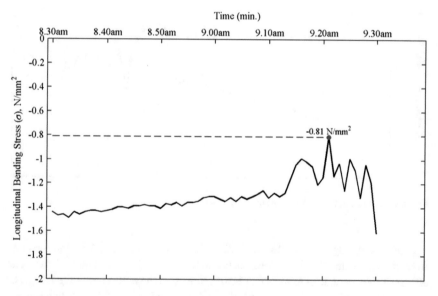

Fig. 4.18 LB$_{stress}$ at web (mid-span)

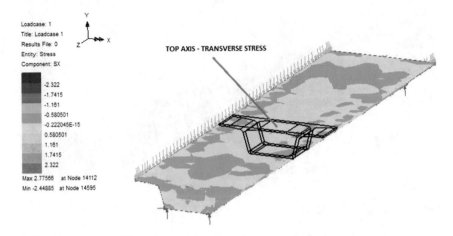

Fig. 4.19 TB$_{stress}$ under LC1 (Top)

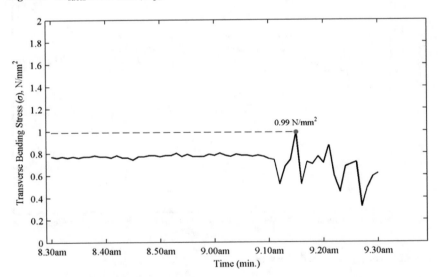

Fig. 4.20 TB$_{stress}$ at top deck (mid-span)

stresses and strains, it can be concluded that the selected single span has responded to the specific load cases and subsequent unloading with measured behaviours and values that are very similar to the predicted value. Furthermore, data collected during this load test measurement was useful to assess the precast SBG's behaviour as well as to evaluate the performance of the bridge under short-term observation.

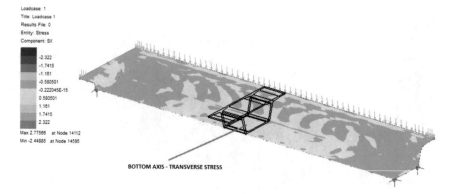

Fig. 4.21 TB$_{stress}$ under LC1 (Bottom)

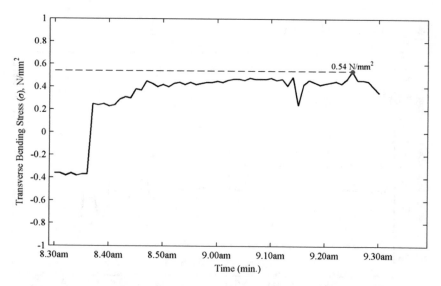

Fig. 4.22 TB$_{stress}$ at bottom deck (mid-span)

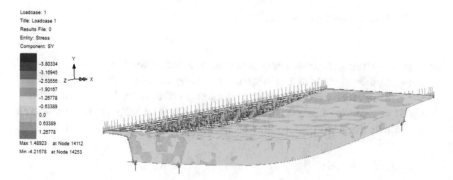

Fig. 4.23 VB$_{stress}$ under LC1 (Web)

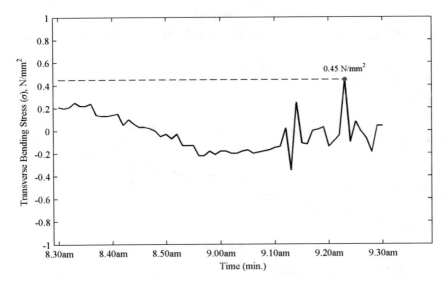

Fig. 4.24 VB$_{stress}$ at web (mid-span)

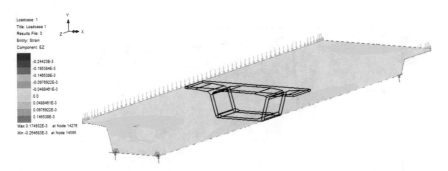

Fig. 4.25 LB$_{strain}$ under LC1

4.4   Chapter Outcome Summary

Linear analysis of the precast SBG bridge cross-sections namely trapezoidal box girder of different transversal slopes (0, 2.0, 2.5 and 3) subjected to static load test and comparison of all transversal slopes was done in term of elastic behaviour, vertical deflection, vertical bending stress, longitudinal bending stress, transverse bending stress and strain at top, bottom and web deck cross-section at mid-span area have been presented. The detailed study carried out using LUSAS software has clearly

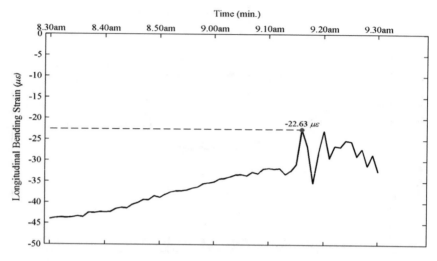

Fig. 4.26 LB$_{strain}$ at top deck (mid-span)

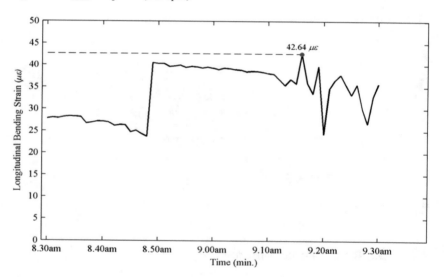

Fig. 4.27 LB$_{strain}$ at bottom deck (mid-span)

brought out the effectiveness of 3-D solid elements for analysis of segmental box girder bridges. The FEM modelling has been validated by experimental works. It can be concluded from the present study that the simple beam theory or traditional method is a crude approximation for analysis of segmental bridge sections.

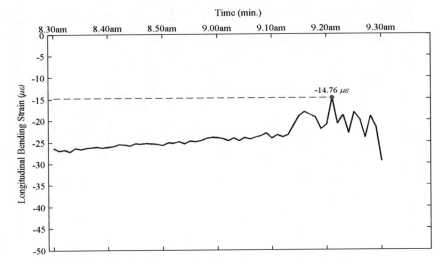

Fig. 4.28 LB$_{strain}$ at web (mid-span)

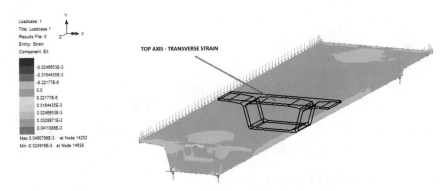

Fig. 4.29 TB$_{strain}$ under LC1 (Top)

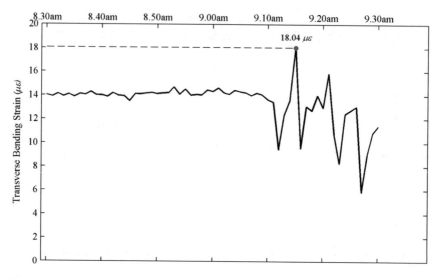

Fig. 4.30 TB$_{strain}$ at top deck (mid-span)

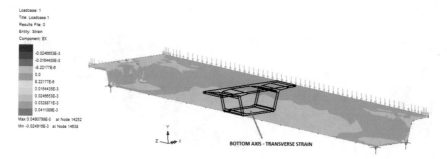

Fig. 4.31 TB$_{strain}$ under LC1 (Bottom)

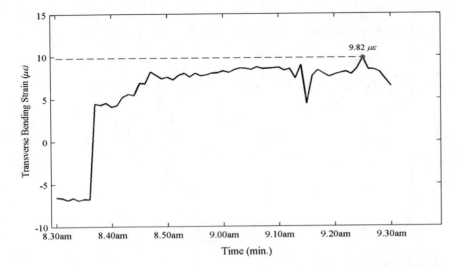

Fig. 4.32   TB$_{strain}$ at bottom deck (mid-span)

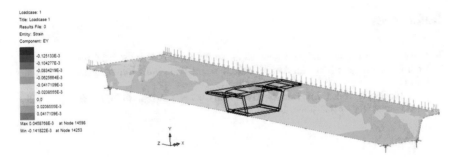

Fig. 4.33   VB$_{strain}$ under LC1 (Web)

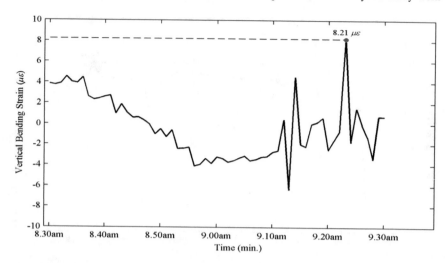

Fig. 4.34 VB$_{strain}$ at web deck (mid-span)

References

1. Department of Physics and Astronomy Appalachian State University. Retrieved 2 Nov 2014, from Error Analysis [Online]. Available from World Wide Web: http://physics.appstate.edu/ (2014)
2. W.H. Mosley, R. Hulse, J.H. Bungey, *Reinforced Concrete Design: to Eurocode 2* (Macmillan International Higher Education, 2012)

Printed in the United States
By Bookmasters